Iva Černá

Partially dehydrated digestate from biogas plant

Iva Černá

Partially dehydrated digestate from biogas plant

Properties and use based on its sorption characteristics

LAP LAMBERT Academic Publishing

Impressum / Imprint

Bibliografische Information der Deutschen Nationalbibliothek: Die Deutsche Nationalbibliothek verzeichnet diese Publikation in der Deutschen Nationalbibliografie; detaillierte bibliografische Daten sind im Internet über http://dnb.d-nb.de abrufbar.

Alle in diesem Buch genannten Marken und Produktnamen unterliegen warenzeichen-, marken- oder patentrechtlichem Schutz bzw. sind Warenzeichen oder eingetragene Warenzeichen der jeweiligen Inhaber. Die Wiedergabe von Marken, Produktnamen, Gebrauchsnamen, Handelsnamen, Warenbezeichnungen u.s.w. in diesem Werk berechtigt auch ohne besondere Kennzeichnung nicht zu der Annahme, dass solche Namen im Sinne der Warenzeichen- und Markenschutzgesetzgebung als frei zu betrachten wären und daher von jedermann benutzt werden dürften.

Bibliographic information published by the Deutsche Nationalbibliothek: The Deutsche Nationalbibliothek lists this publication in the Deutsche Nationalbibliografie; detailed bibliographic data are available in the Internet at http://dnb.d-nb.de.

Any brand names and product names mentioned in this book are subject to trademark, brand or patent protection and are trademarks or registered trademarks of their respective holders. The use of brand names, product names, common names, trade names, product descriptions etc. even without a particular marking in this work is in no way to be construed to mean that such names may be regarded as unrestricted in respect of trademark and brand protection legislation and could thus be used by anyone.

Coverbild / Cover image: www.ingimage.com

Verlag / Publisher:
LAP LAMBERT Academic Publishing
ist ein Imprint der / is a trademark of
OmniScriptum GmbH & Co. KG
Heinrich-Böcking-Str. 6-8, 66121 Saarbrücken, Deutschland / Germany
Email: info@lap-publishing.com

Herstellung: siehe letzte Seite /
Printed at: see last page
ISBN: 978-3-659-69651-0

TABLE OF CONTENT

1. PREFACE

Digestate is beside to the biogas next product of the anaerobic digestion process in biogas plant. It is traditionally used for agricultural purposes, especially as liquid fertilizer. Though the digestate contains important nutrients and is a valuable mineral and organic fertilizer, it is liquid product which has to be utilized immediately or it has to be transported, stored or disposed. Transport and spreading costs can exceed the value of the fertilizer it same, so there are efforts to use this product by another way.

Very interesting became a possibility to dry separated solid part of digestate and compress it to the form of briquettes or pellets. By this way has raised new possibilities of its utilization, like to use it as animal bedding or as a solid fuel. Because these ways are relatively new, the properties of granulated digestate has not been properly described yet.

This book is focused on this problematic, especially to describe the physico-mechanical properties of digestate briquettes in related to unknown page of its sorption properties. Three targets have been stated namely: to increase awareness of basic, mainly mechanical, properties of the compressed digestate; to determine by laboratory experiments sorption potential of the digestate briquettes, in comparison with a different material; and last is based on gained properties and measured values changes in digestate briquettes properties depended on water uptake.

To describe its properties was used basic laboratory analyses and experiments. Three types of experiments were investigated; granulated digestate was exposed to sorption in non-limited, partially limited and completely limited conditions. By these methods were observed changes in gravimetric values. All measured values were compared to values of briquettes of different materials.

Along with the acquired knowledge of digestate in the form of briquette, may be better to propose other use of digestate such as organic fertilizer or as a means to modify the physical properties of the soil.

2. INTRODUCTION

2.1 Anaerobic digestion (AD)

2.1.1 Basic information on AD

2.1.1.1 History and current situation

The technology for anaerobic digestion was known for centuries before Christ, in the time of Assyrians and Persians rule (Bond, et al., 2011), (Kumar, 2012). This knowledge is one of the oldest skills, but it became popular in 19th century as renewable technology in time of faster growth of livestock industries, because with increase of animal waste, disposal requirements also have increased. At the beginning AD was used just for treatment of sludge, while developing countries such as China was innovating skills and used the digestion to energy production and sanitation purposes. Under the pressure of energy prices rising, AD stations were diversified to large scales. In Europe the anaerobic digestion market was developed because of higher energy prices and strict environmental regulations (Monnet, 2003). Because AD technology has been well studied, now is successfully used in both large even small scales, mainly for energy production then for treatment of different kinds of wastes as sewage, agricultural, industrial and municipal degradable biomass (Gong, et al., 2010), (Bond, et al., 2011). The European area counts up around thousands of biogas plants and is leader in the world, with 91%. Main European countries AD operators are Germany (35%), Denmark (16%) Austria (8%), Sweden and Switzerland (Chaudhary, 2008) and (Seadi, et al., 2008) - over 600 simple farm-scale reactors run in Europe (Monnet, 2003). Europe or USA (2%) is not the only big biogas operator; also other parts of the world, developing countries, are using this technology, e.g. Asian countries (7%) China with app. 18 million household stations in 2006 or India with 5 million small-scale BP and Vietnam as well, these technologies are very simple but useful (Seadi, et al., 2008).

2.1.1.2 AD process

Method of whole digestion process can be split into four stages: pre-treatment, digestion itself, gas upgrading and digestate post-treatment, see Figure 1. First point, pre-treatment, is influenced by feedstock composition. Some substrates going into the digestion process have to be mixed and another have to be sorted and slashed, such as MSW. As a result digestate levels increase in the reactor. Many types of digesters are known with different parameters. Process of digestion is dependent on solid content, than we distinguish wet or dry digestion. Produced biogas has to be upgraded, because of impurities in the gas can damage engines or boilers. Also hydrogen sulphide, water vapour and carbon dioxide, if the gas is going to be used as natural gas or fuel, have to be removed.

A more improved method of AD is Co-digestion; a type of anaerobic digestion where two different categories of biomasses are mixed. Examples of such mixtures are; animal manures and slurries from slaughterhouse wastes with other organic matter like glycerin,

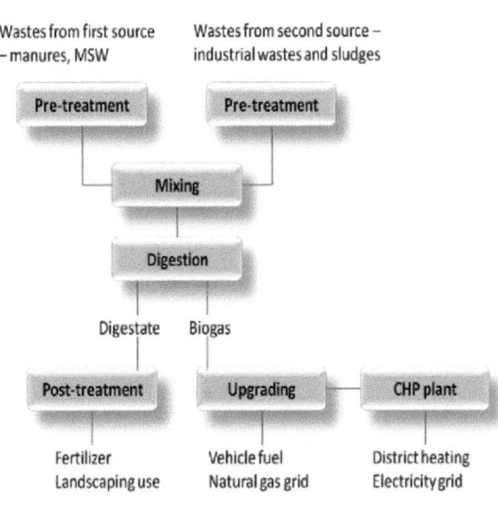

Figure 1: Upgrading digestate stages (Monnet, 2003)

energy crops or silage. The benefits of these mixtures include; improved biogas yield per m^3 of reactor, the dilution of potential toxic compounds, easier handling of slurries when mixed with fluid manure, more efficient results of digestion. Finally Co-digestion can improve the NPK ratio of a by-product and favor synergistic growth of microorganisms (Monnet, 2003), (Alburquerque, et al., 2012).

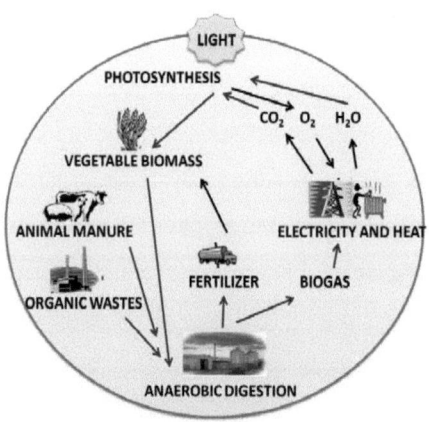

The whole practice of AD is for carbon dioxide (CO2) to be a neutral energy creator in a nutrient closed cycle with the possibility of pathogen inactivation, and weed seed and germ elimination (Deublein, et al., 2008), (Seadi, et al., 2008), see Figure 2. This process commonly takes place in natural conditions where contributes to an annual increase of methane level in the

Figure 2: CO2 and nutrients closed cycle (Seadi, et al., 2008)

atmosphere by 590-800 million tons (Bond, et al., 2011). The essence of digestion is the microbial decomposition of organic matter in anaerobic conditions; it means AD is catabolic process with absence of free molecular oxygen O_2 (Seadi, 2001), (Gerardi, 2003). Due to the metabolism of various groups of microorganisms, a mixture of gas is released and the residual substrate rests the digestate (Seadi, 2001).

2.1.1.3 Biogas

Methane (CH_4) is the main part of biogas, energetically usable (flammable with >45% of CH_4) by-product of anaerobic digestion (Deublein, et al., 2008); (Seadi, 2001). Biogas components (see Table 1) are variable according plant and substrate use. These factors are regularly controlled (Deublein, et al., 2008). Biogas from AD is mainly composed from methane and carbon dioxide, then smaller quantity of hydrogen sulphide is contained and ammonia as well. Next components are present in trace amounts, it is hydrogen, nitrogen, and carbon monoxide, saturated or halogenated carbohydrates and oxygen. Gas is also saturated with water vapor, which is important to be removed, because of problems in gas nozzles. With water remove the large proportion of hydrogen sulphide is also detached and it is preferable because of formation sulphurous acids with corrosive effects (Monnet, 2003). Damaging effects like decreasing of calorific value, imperfect combustion and/or damage of the engine, can be caused by other impurities like dust particles and siloxanes etc. (Deublein, et al., 2008).

Table 1: Composition of biogas (Seadi, 2008)

Compound	Chemical symbol	Content (Vol.-%)
Methane	CH_4	50-75
Carbon dioxide	CO_2	25-45
Water vapor	H_2O	2 (20°C) -7 (40°C)
Oxygen	O_2	<2
Nitrogen	N_2	<2
Ammonia	NH_3	<1
Hydrogen	H_2	<1
Hydrogen sulphide	H_2S	<1

Artificial biogas is different from the natural gas (see Table 2), but has comparable composition with landfill biogas. The composition of the biogas is varying according digested waste composition. Calorific value of natural gas is 36.14 MJ/m^3 and of biogas 21.48 MJ/m^3 (Monnet, 2003).

Table 2: Comparison of composition CH_4 containing gases (Kumar, 2012); (Monnet, 2003)

Component	Natural gas (%)	Biogas (%)	Landfill gas (%)
CH_4	85-91	50-80	45-58
CO_2	0.89	20-45	32-45
C_2H_6	2.85	-	-
C_3H_8	0.37	-	-
C_4H_{10}	0.14	-	-
N_2	14.32	0-2	0-3
O_2	<0.5	-	-

H₂S	<0.5	0-1.5	10-200 ppm
NH₃	-	0-0.45	-
VOC	-	-	0.25-0.50
H₂	-	-	Trace
CO	-	-	Trace

Biogas utilization is similar to natural gas utilization with the same quality standards. Examples of biogas use: heating boilers used in the plant or for industrial purposes, combined heat and power units CHP, fuel for vehicles or fuel cell which generates power in electricity form by combining fuel and oxygen in electrochemical reaction (Monnet, 2003). For proper and safe biogas exploitation there are some technologies for its upgrading. For enhance energy of the biogas there are four techniques for CO_2 remove: water scrubbing, polyethylene glycol sieves and membrane separation. As was mentioned to avoid corrosion hydrogen sulphide has to be remove as well, the most widespread techniques are: air/oxygen dosing to digester gas, iron chloride dosing to digested slurry, activated carbon, water scrubbing, NaOH scrubbing and/or iron oxide. Desulphurization can be managed also by micro-organisms from the family *Thiobacillus* with addition of oxygen. Hydrocarbons causing corrosion can be filtered by use of pressurized tube exchanger with activated carbon filter; all molecules ot gas are going through the filter thus large are absorbed. By the absorption on a liquid medium siloxanes are removed (Monnet, 2003).

It was estimated by scientists and experts, that only a small fraction of biogas production potential is used nowadays, there exist possibility to increase actual production considerably. (Seadi, et al., 2008)

2.1.2 Principles of AD

The biomass used for fermentation contains carbohydrates, proteins, lipids and other organic compounds and before using this material as feedstock to the AD, nutritional values of these components have to be taken into consideration to be appropriate for gas formation. (Deublein, et al., 2008) Chemically bonded energy is locked in biogas;

its formation results from a four level continuous break down of the initial material. As previously mentioned different groups of micro-organisms are main actors in this process, decomposing the product of the previous level (Seadi, et al., 2008).

AD four phases are: hydrolysis, acidogenesis, acetogenesis and methanogenesis, see Figure 3 (Chaudhary, 2008). All levels are parallel in time even in space of digesting reactor. Total speed of whole process is given by the slowest reaction of the chain, e.g. in decomposition process of crop compounds containing cellulose, hemi-cellulose and lignin, influencing level is hydrolysis (Seadi, et al., 2008). And if the first phase is inhibited, the substrate for next phases is limited and the production of methane will decrease. It is the same with other phases of AD, with unwanted results

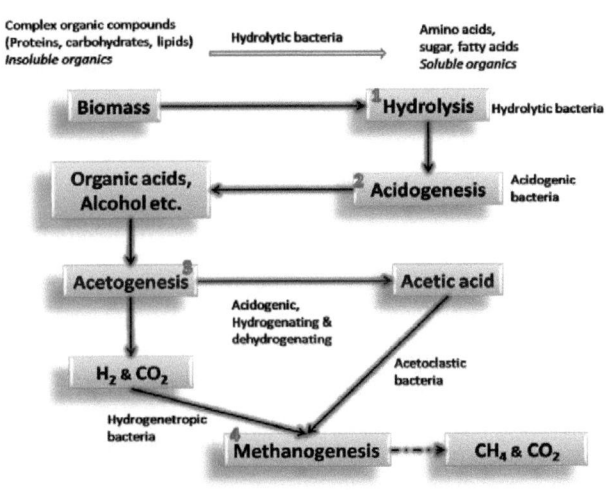

Figure 3: Anaerobic digestion process (Chaudhary, 2008)

like accumulation of acids, losing of alkalinity and pH decrease thus methane production decrease (Gerardi, 2003). Individual phases are described in detail below:

Hydrolysis: hypothetically the first stage of digestion where complex insoluble organic polymers are decomposed. High weight complexes such as lipids, polysaccharides, proteins and nucleic acids are converted into smaller soluble molecules of glucose, glycerol, purines, and pyridines. Hydrolytic micro-organisms or facultative anaerobes and anaerobes are excreting hydrolytic enzymes, which are breaking simple unique bonds in soluble compounds (Gerardi, 2003); (Appels, et al., 2008); (Seadi, et al., 2008). Term hydrolysis is composed from -lyses means splitting bonds and hydro- means with water (solubility).

An example of cellulose digestion; it is allowed thanks hydrolytic bacteria *Cellulomonas* by enzyme cellulase, which can break bonds between unique mers of glucoses. Even though glucose is soluble in water, when present in cellulose, the polymer makes the unique bonds insoluble (Gerardi, 2003). Next enzymes essential for digestion are lipase splitting lipids bonds, already mentioned cellulase, cellobiase, xylanase and amylase splitting polysaccharides bonds and protein bonds splitting, protease. The products resulting from the hydrolysis stage are metabolized by the other microorganism involved and are decomposed further (Seadi, et al., 2008).

Acidogenesis: in this stage compounds made in the first step of digestion are converted by large diversity of acidogenic facultative anaerobes and anaerobes thru fermentative processes into methanogenic substrates. Simple sugars, amino acids and fatty acids are converted into carbon dioxide (CO_2), hydrogen (H_2), alcohols (-OH), volatile fatty acids (VFO), some organic-nitrogen and organic-sulfur compounds and the most important acetate (Gerardi, 2003); (Seadi, et al., 2008).

Acetogenesis: during the acetogenesis process all products which cannot be directly converted into methane by methane-forming bacteria are converted into methanogenic substrates such as; acetate, formate, methanol, and methylamine (Seadi, et al., 2008). Compounds like acetate, formate, methanol, and methylamine, it means organic acids, some alcohols and organic-nitrogen compound are used by methane-bacteria directly (Gerardi, 2003). Volatile fatty acids (butyrate, propionate) and alcohols like ethanol can be used indirectly, thru degradation into acetate, hydrogen and carbon dioxide by fermentative bacteria (Gerardi, 2003); (Seadi, et al., 2008). This conversion has to be controlled because of increasing hydrogen partial pressure, inhibiting the metabolism of acetogenic or acetate-forming bacteria (Appels, et al., 2008); (Seadi, et al., 2008).

Methanogenesis: the last stage methanogenesis is parallel with acetogenesis stage; symbiosis of two groups of microorganisms is common. First group of Methanogenic bacteria is converting acetate (70%) and the second group uses hydrogen (H_2) as an electron donor and carbon dioxide (CO_2) as an acceptor (remaining 30%) to creation of methane and carbon dioxide (Appels, et al., 2008); (Seadi, et al., 2008), shown below:

(Acetoclastic methanogenesis) Acetic acid $\xrightarrow{\text{methanogenic bacteria}}$ *methane + carbon dioxide*

(Hydrogenotrophic methanogenesis) Hydrogen + carbon dioxide $\xrightarrow{\text{methanogenic bacteria}}$ *methane + water*

Methanogenesis is the slowest stages in the whole process of digestion. This biochemical reaction is influenced by operational parameters like feedstock composition, rate and volume, temperature in reactor, pH and others. Improper compliance and settings of these parameters can cause termination of methane production (Seadi, et al., 2008).

2.1.3 Parameters of AD

Durability of involved particular bacteria and the efficiency of whole technology are radically influenced by critical parameters that must be followed (Seadi, et al., 2008). We can divide parameters to operational and AD process parameters, which are:

Temperature and Hydraulic Retention Time – temperature is important to physicochemical properties of substrate to be digested and for metabolism of bacteria and their growth rate (Appels, et al., 2008). There are three temperature range possibilities, under 25°C is psychrophilic temperature range, 25°C - 45°C is mesophilic and 45°C - 70°C is thermophilic digestion (Seadi, et al., 2008). The best possible temperature varies depending on feedstock composition and type of digestate reactor, but it should be maintained without larger fluctuations (Monnet, 2003). Thermophilic bacteria are more sensitive for temperature changes entail they need longer adaptation time for new temperature. Mesophilic bacteria are more tolerant in temperature changes. Fluctuations can be caused for example by: adding of new substrate that has different temperature than current one, insufficient insulation, inaccurate heating system or mixing, extreme outdoor temperatures or failure of power-trains (Seadi, et al., 2008). Heating can be ensured by external heating system or by internal heat exchangers by steam injection (Appels, et al., 2008).

Temperature and the hydraulic retention time (HTR) guarantee a stabilized digestate (Seadi, 2001); it is important to next digestate utilization. The relation between temperature and the HTR is shown in Table 3:

Table 3: Anaerobic digestion temperature stages (Seadi, et al., 2008)

Thermal stage	Process temperatures	Minimum retention time
Psychrophilic	< 20 °C	70 to 80 days
Mesophilic	30 to 42 °C	30 to 40 days
Thermophilic	43 to 55 °C	15 to 20 days

Thermophilic temperature is preferred as an operational temperature, because is more efficient in terms of retention time, loading rate and gas creation (Monnet, 2003); (Seadi, et al., 2008). Advantages of thermophilic digestion are: e.g. pathogens destruction is more effective, growth rate of methanogenic bacteria is higher, retention time is shortened and thus makes digestion faster (faster chemical reactions) and more efficient, better degradation (higher solubility of OM), digestibility availability and utilization of substances etc. Of course there are some disadvantages like higher energy demand, higher risk of ammonia inhibition or larger imbalance (Seadi, et al., 2008).

Hydraulic retention time (HRT) of the fluid biomass in the digester is time needed to reach the complete decomposition of organic matter. Measured in days correlates with digestion temperature and waste composition (Seadi, 2001); (Seadi, et al., 2008). Minimum retention time is min. time interval, measured in hours, for which any part of digested substrate will be inside a continuous reactor (Seadi, 2001). HRT shorter than 5 days is insufficient causing washout of methanogenic bacteria. In 5-8 days decomposition is still incomplete, especially of lipids. Stable digestion may occur after 8-10 days with increasing of VFA and breakdown of lipid bonds (Appels, et al., 2008). In relation to temperature, retention time for mesophilic digestion lasts of around 30 days and for thermophilic around 15 days (Monnet, 2003); (Seadi, et al., 2008).

pH-value, VFA and ammonia: pH value expresses acidity vs. alkalinity rate of substrate. It is affecting growth of microorganisms and dissociation of ammonia, sulphide and/or organic acids in the substrate (Seadi, et al., 2008). Each group of microorganisms requires different pH range (Appels, et al., 2008). Methanogenic bacteria are extremely pH sensitive and experiences show that methane production is going in relatively narrow interval of 5.5-8.5. For most methanogens the optimum pH value is between 7.0 and 8.0 (Seadi, et al., 2008). Changes in pH value can cause inhibition of one of stages of AD and pH below 6.0 can even be toxic for methane-forming bacteria (Monnet, 2003). There are some controlling limits for pH value, e. g. presence of ammonia and VFA, then bicarbonate buffer system (70 mg $CaCO_3/l$), where pH value is dependent on concentration of alkaline and acid components and partial CO_2 pressure.

VFA and their concentration in the substrate are important for stability of AD. Acetate, butyrate, propionate and lactate are formed during acidogenesis and their accumulation, causing AD instability, leads to a drop in pH. As well as VFA, free ammonia in digester can cause AD inhibition. All changes in ammonia concentration are connected with pH value and temperature. With increasing pH and temperature, fraction of free ammonia is also increased, leading to inhibition of AD and VFA concentrations (Appels, et al., 2008); (Seadi, et al., 2008).

Solids content, nutrients, C:N ratio: ranges of solid content are LS – low solids with less than 10% of Total solids (TS), MS – medium solids with 15-20% and HS – high solid with 22-40% of TS. According content of solid matter, AD process can be divided into wet or dry digestion (Monnet, 2003). The next important components for the proper function of digestion, mainly for growth and survival of microorganisms, are micronutrients – iron, nickel, selenium, cobalt, molybdenum or tungsten. Even macronutrients are essential for proper microbial functioning; important nutrients are carbon (C), nitrogen (N), sulphur (S) and phosphor (P) with optimal ratio of C:N:P:S is 600:15:5:1 (Seadi, et al., 2008). C:N ratio present relationship between C and N content in organic material and the optimal range is 20:1-30:1 (Monnet, 2003). But because N can be bounded in lignin structures, C:N ratio is just an indicator (Deublein, et al., 2008). To keep optimal C/N it is advised to mix wastes with different values see example in Table 4. (Monnet, 2003).

Table 4: C/N ratio for different types of feedstock (Monnet, 2003)

Feedstock	C/N ratio
Pig slurry	3-10
Cow slurry	6-20
Chicken slurry	3-10

Operational parameters and division of digesters: operational parameters vary in range of complexity from small scale cylindrical reactor without moving parts to fully large scale automated industrial facilities. Design includes vertical and horizontal orientations, batch or continuous flow, wet or dry technology, single step or multiple, etc. All parts are designed to optimize the process as much as possible, according to geographic location, types of feedstock available and others factors (Zaher, et al., 2007). Common digesters used mainly in small scale, in developing countries. Very frequently used are fixed dome digesters (Chinese type), floating cover digester (Indian type) or balloon or tube digester see Figure 4 (Bond, et al., 2011), (Plöch, et al., 2006). All possible types of digestion are designed in Figure 5. It is classification of anaerobic digestion by operational criteria. (Chaudhary, 2008)

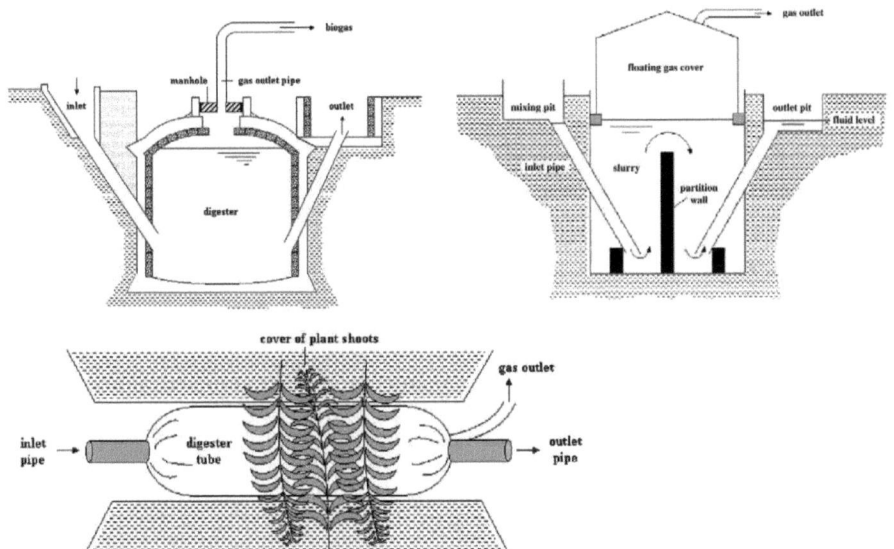

Figure 5: Anaerobic digesters, Chinese, Indian and PVC type (Bond, et al., 2011)

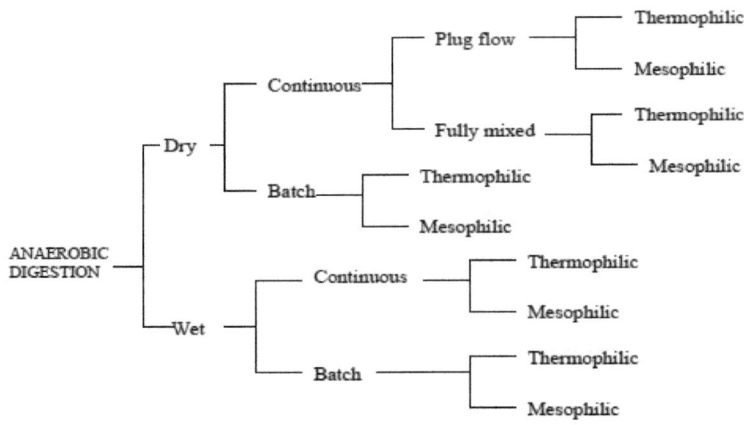

Figure 4: Classification of AD (Chaudhary, 2008)

Feedstock: substrates used for anaerobic decomposing of material varies, the most common biomass materials suitable for AD are for example sewage sludge, organic farm wastes and municipal wastes, green/botanical wastes and organic industrial and commercial wastes. By mixing different types of feedstock, co-digestion can have an impact on environmental and economic aspects. (Chaudhary, 2008). Feedstock is basically divided to agricultural and non-agricultural waste. Examples of wastes used in anaerobic digestion: Agricultural waste and agricultural vegetation waste include different animal manures, residues after crop cultivation, including mushroom medium residues, gardening residues and animal bedding derived from biomaterial (straw, paper, hog fuel, wood chips, bark, shavings or sawdust). Next acceptable feedstock used is waste products from animal feeds, cooking oil and food residues from restaurants, milk organic by-products from biodiesel facilities, plant matter etc. Materials which are limited or not accessible to biogas fermentation are: bio-solids, fish wastes, hatchery wastes, pet food and residues, poultry wastes or red-meat, hazardous material, catering waste from international transport, mortalities that have died to infectious diseases, organic wastes with content of fuels, plastic components or organic waste with volatile organic compounds.

To ensure quality biogas production, it is recommended to sample each applicable feedstock type, to analyze physical characteristics (moisture, ash content, pH, C/N and others), chemical characteristics (nutrient content) and mainly heavy metals content. (BC Ministry of Environment, 2010).

2.1.4 Treatment

Digestate as a result of the AD is sometimes not fully stabilized, so residual biodegradability and organic elements like ligno-cellulosic compounds can be presented. Residues of digestion should be subjected to an appropriate treatment, the post-treatment. It is because to ensure suitability to agricultural use, mainly safety. The adequate post-treatment method could be composting, since it can stabilize organic matter and reduce phyto-toxicity and even advance the humic potential (Teglia, et al., 2011). Main treatment operations are illustrated in Figure 6 and 7.

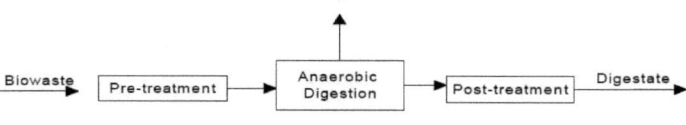

Figure 6: Waste processing (Seadi, 2001)

Unit processes	Reusable products	Standards or criteria
PRE-TREATMENT - Magnetic separation - Size reduction (drum or shredder) - Pulping with gravity separation - Drum screening - Pasteurization	- Ferrous metals - Heavy inerts reused as construction material - Coarse fraction, plastics	- Organic impurities - Combination of paper, cardboard and bags - Germs die off
DIGESTION - Hydrolysis - Methanogenesis - Biogas utilization	- Biogas - Electricity Heat (steam)	- nitrogen and sulfur contents - 150 - 300 kWh elec /ton 250 - 500 kWh heat /ton
POST-TREATMENT - Mechanical dewatering - Aerobic stabilization or Biological dewatering - Water treatment - Biological dewatering - Wet separation	- Compost - Water - Sand, Fibres (peat), Sludge	- soil amendments - Water treatment load - Disposal regulations - Organic impurities

Figure 7: Treatment processes (Zaher, et al., 2007)

2.1.5 Benefits and limitations

Process of AD has both benefits and drawbacks (Seadi, 2001). The main advantages of controlled anaerobic digestion are considered to be: low levels of emitting green house gases if the process is well-managed and the methane production is maximized (Monnet, 2003), AD is source of energy both electrical and thermal with a closed carbon cycle. Heat energy can be recovered from waste heat or produced by burning of biogas. Then biogas can be converted to the vehicle fuel, like a CNG or LNG (EPA, 2012). Because renewable resources are used as a feedstock, the demand for fossil fuels is decreased and by using the digestate as a fertilizer, industrial fertilizers utilization and manufacturing is inhibited. In comparison to untreated and unmanaged disposal to animal manures, AD as an integrated management system reduces the probability of soil and water pollution. By proper treatment, the prevalence of odors and weed seeds are depleted, thus need for herbicides and other weed control are less essential (Monnet, 2003). Another example of how to utilize and profit from digestate production is to use it as an animal bedding, or use it like a solid fuel – to combust it (EPA, 2012). In financial point of view all AD by-product are converted to saleable products like biogas, soil amendment, fertilizer. This ensures economic viability of farm (Monnet, 2003).

There are also some risks in AD utilization and some potential negative influence on the environment. Biogas production is very variable and must be treated, before it can be used, which increases costs, in addition to the high capital and operational costs. In different countries this can create difficulties with local policies and interconnection challenges and in rural areas even capacity issues. Costs can be also influenced by distances between production of the feedstock, storage facilities and digester. These distances and the traffic movement should be minimized (EPA, 2012) and (Monnet, 2003). Even risks to human health can occur, like pathogenic content of the feedstock, which can be decreased by suitable plant design and feedstock handling. In most cases it has visual impact to surroundings. Even risk of fire and explosion is possible (Monnet, 2003).

2.2 Digestate

2.2.1 Organic matter and fertilizers

2.2.1.1   Organic material and biomass

According the definition biomass is organic, carbon based material, derived from both animal and plant living, or recently dead material (GTOS, 2009). Agriculture mainly uses residues from animal or crop production. These residues are rich for most major plant nutrients and organic matter. By recycling of biomass the nutrient value is given back into the soil fertility which can save on using industrial fertilizers (DEFRA, 2010). Organic residues are used for agricultural land in specified doses and in appropriate period of application (Teglia, et al., 2011). In prevalence agricultural organic wastes are on form of particularly solids manures. In the soil, these manures contribute to improvement of water holding capacity, drought resistance and structural stability and the biological activity (DEFRA, 2010). In current situation in Czech Republic lots of farmers are breaking the system that was functional from 17th century. The system used proper planted crops, proper crop rotation by using clover crops and fertilized the soil with valuable organic manures – fertilizers. The essence is to supplying agricultural soil by both nutrients and organic matter. By the decrease of cattle farming, even the clover plants were excluded from crop rotation. The main characteristics of these plants are aeration of the soil thanks to a deeper root system, releasing nutrients like Ca, Mg or P from a deeper level in the soil. After plowing, the root system becomes a big source of material for the production of soil humus matter and the reclamation of the soil. The main benefit is the bonding of air nitrogen into the soil thanks to the bacteria *Rhizobium sp.* The next positive aspect is in the dense coverage of the soil, thus to lessen the effect of soil erosion agents. The benefit of using other organic fertilizers, like animal manure, in liquid form is inter alia giving considerable amount of water into the soil (Marešová, et al., 2012). Because of the tendency to enrich the soil using mainly industrial fertilizers, there are trials using different organic materials, like the digestate and others (FAO, 2005).

2.2.1.2 *Decomposition of organic matter*

Fresh residues, according to its origin, consist of recently dead organisms, both micro- and macro-microorganisms, including insects, earthworms and others. In the soil all organic compounds undergo process of decomposition. It is the process of braking down and transformation of organic molecules into simpler organic and inorganic parts. With addition of decaying organic matter the biological activity and the carbon cycling process starts in the soil, see Figure 8. Quality of organic matter, with other factors, influences the speed of decomposition. The process of breaking down of organic carbon compounds makes the humus, which influences soil properties and nutrients balances. The more stable humus enables increase of stored water and C from the atmosphere. Humus

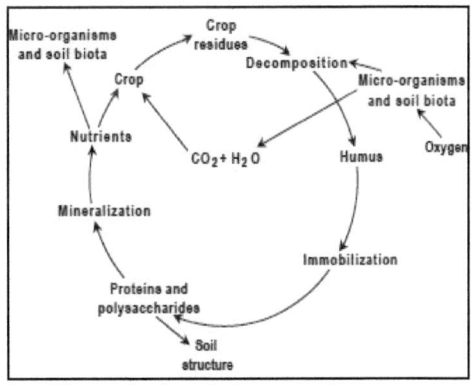

Figure 8: Decomposition cycle (FAO, 2005)

remains in the soil relatively long time, because it is difficult for micro-organisms to use it as an energy source. Other functions of humus are: it improves fertilizer efficiency, keeps longlife of nitrogen, improves nutrients uptake (particularly of P and Ca), stimulates of beneficial soil life, provides magnified nutrition for reduced disease, insect and frost impact and manage salinity – humates 'buffer' plants from sodium excess (FAO, 2005).

2.2.1.3 Soil moisture and water saturation

With increasing precipitations levels of soil organic matter grows. Soil biological activity is dependent on both air and moisture. The optimal biota activity arises, when around 60% water - resulting in a filled pore space. However soil decaying microorganisms need oxygen as well and in case of overfilled pore space, poor aeration occurs and with it even reduction of mineralization, inactivation and/or death of organisms. Other processes became active, anaerobic, causing damages of roots and diseases (FAO, 2005).

2.2.1.4 Human interventions that influence soil organic matter

By human various activities the organic material level and biological activity in the soil can be decreased. The practice of reducing organic matter in the soil by human intervention can influence the equilibrium of soil organisms. For example repeating tillage or burning of vegetation decrease microenvironment. Thus it destroys the soil structure and it no longer contains the organisms required to decompose the organic compounds. Soil is easily damaged by rain, wind and sun leading to rainwater runoff and soil erosion, thus removing food for organisms of the topsoil.

Here are some factors that decrease the soil organic material in an open cycle system (Figure 9): decrease in biomass production – means replacement of perennial vegetation, and mixed vegetation with monoculture of crops

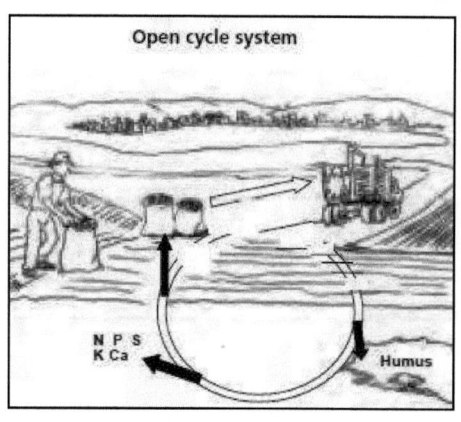

Figure 9: Open cycle system (FAO, 2005)

and pastures, high harvest index, use of bare fallow; decrease in organic matter supply - burning of natural vegetation and crop residues, overgrazing, Removal of crop residues; increased decomposition rates - tillage practices, drainage, fertilizer and pesticide use. Compare to aforementioned practices there are some that increase the organic matter in the soil, thus providing habitat and food for soil organisms, which help build soil structure and porosity, provide nutrients and improve water holding capacity of the soil. Ways to increase OM content are: use of cover crops, compost, and proper crop rotation, perennial forage crops, reduced or zero tillage and/or agroforestry system (FAO, 2005).

2.2.1.5   Nutrient requirements of crops

According relative impotence nutrients are divided to macro and micro, in addition to carbon, hydrogen and oxygen. Macronutrients are nitrogen (N), magnesium (Mg), phosphorus (P), calcium (Ca), potassium (K) and sulphur (S) and are required in relatively large amounts. Micronutrients, required in smaller amount, than macronutrients are: iron (Fe), copper (Cu), manganese (Mn), zinc (Zn), boron (B), molybdenum (Mo) and chlorine (Cl). Deficiency of any of macro- or micronutrient can limit growth of plant and thus decreases yield. There is essential to ensure an optimum supply of all nutrients. Crops achieve nutrients from several sources, like from mineralization of soil organic matter, deposition from the atmosphere, biological nitrogen fixation, weather of soil minerals, application of organic manures, application of manufactured fertilizers, other materials added to land e.g. soil conditioners. Nutrients presented in industrial fertilizers are expressed e.g. in oxide form: phosphate (P_2O_5), sulphur, magnesium and sodium (SO_3, MgO and Na_2O) and potassium as potash (K_2O). Right utilization consists in adequate timing and applying correct amount of the fertilizer. Crop demand for nutrients is dependent on the growth season and phase of growth. The most crucial stage is during the development of leaves and root system, the early stage. Wrong timing can reduce crop quality and can cause problems, like increase of foliar pathogens. Other elements contained in plants matter are cobalt (Co), nickel (Ni), selenium (Se), silicon (Si) and sodium (Na) are normally available in the soil for uptake by plant root system and are essential for animal nutrition. These, absorbed in different forms, have special functions within the plant and can cause different deficiency or even toxicity (DEFRA, 2010).

2.2.1.6   Nutrients and fertilizing management

Supply of all mentioned nutrients has to be met, but not exceeded. Requirement varies with species, yield potential and intended use. Nutrients are applied, in sufficient amount, in form of organic manures and inorganic fertilizers if it is needed. There exist recommendation plans for matching right nutrients to be use for fertilizing, using soil Index systems.   In all agricultural production nutrients are matched according crop needs, maximizing economic returns and minimizing nutrients losses (DEFRA, 2010).

Good nutrient management is important for good yields and for a healthy environment. By using fertilizers on farm, human is somehow influencing the environment; the main issues are contamination of surface and ground water, impacting to the quality of the air, contributing to GHG emissions, soil quality etc. Good management of fertilizers use is surely very essential for both profitability of farm and its environment impact. Optimizing the nutrient uptake ensures the minimal excess to the soil and subsequent lost in emission or other forms. Using of mineral fertilizers with organic wastes like slurries, manures, digestate and others helps to

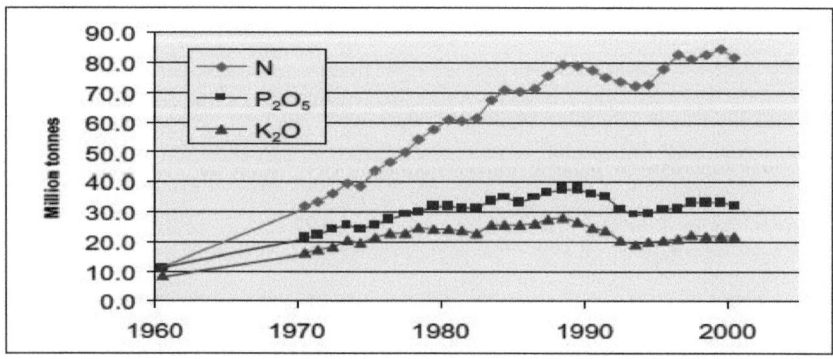

Figure 10: Fertilizers world consumption (Fixen, et al., 2002)

close the 'nutrient gap' (DEFRA, 2010). The fertilizer industry is highly efficient and the world consumption during the years 1960 – 2000 has increased; in case of nitrogen fertilizers by 80% (Figure 10), it has the potential to be utilized to a greater extent. Because animal manure nutrients have a low efficiency level, nitrogen can leak into groundwater or the atmosphere, industrial fertilizer could be used to reduce losses and allow agriculture to utilize manure management better (Fixen, et al., 2002). Production of industrial N fertilizers is very energy-consuming, app. 1.2% of world primary energy. The essential compound of nitrogen fertilizer is ammonia, which production is based on natural gas, unfortunately gasification of coal and heavy oil can occur. The new aim of producing industrial fertilizers is to lessen the environmental impact and decrease fossil energy use by using anaerobic digestion and production of N fertilizers from biogas instead of natural gas. By replacing natural gas with biogas in N fertilizer production, agricultures dependency on fossil fuels will be decreased. Another positive aspect is, that production will contribute to AD digestate creation, used as a fertilizer (Ahlgren, et al., 2010).

2.2.2 Characterization of digestate and Properties

Digestate is produced by biodegradable process of anaerobic digestion, consisting of a mix of microbial biomass and undigested material. Volume of produced digestate is approximately the same as the entering volume of feedstock to AD, but the mass is reduced by app. 15% (Frischmann, 2012).

In present time the digestate is used more frequently, it is not surprising when a common biogas plat of 500kW power efficiency produce more than 10,000 tons of digestate per year some has to be disposed (Kratzeisen, et al., 2010). To assure a suitable and reliable way to manage and recycle these AD residues relevant characteristics of digestates organic matter composition have to be designed. Only a limited amount of data describing different types of digestates is available, concerning the evaluation of digestates stability (Teglia, et al., 2011).

2.2.3 Digestate fertilizer

Digestate has been used as a fertilizer because of organically bound basic nutrients that are mineralized (NPK) and are easily available for plants (Seadi, et al., 2008). Typical values of nutrients are N: 2.3 – 4.2 kg/tonne, P: 0.2 - 1.5 kg/tonne and K: 1.3 - 5.2 kg/tonne; but actual nutrients content is variable according the type of digested feedstock (Frischmann, 2012).

During the digestion mainly carbons, hydrogen and oxygen is consumed and the rest essential plant nutrients remains in the digestate substrate. In comparison to others organic manures, digestate has higher availability of these nutrients, e.g. 25% more accessible NH_4-N and higher pH value (Monnet, 2003). However a large part of nitrogen can be converted, during the digestion, into ammonium and thus lead to phyto-toxic material or emissions (Teglia, et al., 2011). Digestate has lower C:N ratio, that means it has better short-term N- fertilizing effect, and is more homogenous than raw slurry and has an improved N-P balance (Seadi, et al., 2008). There are some differences between statuses of digestate as a fertilizer; most of them are considering digestate as an organic fertilizer, but there is as well an idea, that use of term 'organic' is inaccurate. Kužel et al., 2010 is describing digestate just as light mineral fertilizer, if is in solid form, which is the most useful as an amendment medium for alteration of heavy clay soils. (Kužel, et al., 2010) But the agricultural use is mainly dependent on the digestate quality. There is a tendency to dewater the digestate, meaning that the matter of digestate is separated into fiber and liquor (Monnet, 2003).

Forms

There are two forms in which digestate can exist, according of the nature of feedstock and digestion technology; it means the liquid and solid form. Liquid digestate is mostly directly spread onto the soil (Teglia, et al., 2011). In form of liquid, digestate contains typically 10% of dry matter (Kratzeisen, et al., 2010). Spreading onto the soil, of course, shall be the subject of specific limitations and legislature regulations see chapter Legislative aspects. I case of AD with separation facility, solid/liquid part is divided to produce different by-products for different utilization. The liquid fraction can be used as a fertilizer after treatment procedure or undergo ammonia stripping. The solid fraction is mostly used for amendment of soil conditions or for other purposes. The solid form has the big advantage, the possibility to be economically handled and transported, and can prevent the risk of groundwater pollution. After the treatment of digestate residue (centrifugation, pressing and drying) the final solid fragment is possible to use as amendment of soil.

Liquid form characterization

In comparison to other organic fertilizers, liquid digestate has these properties: total nitrogen content of 0.2 – 1% in dry matter, higher pH (7-8), lower content of carbon and dry matter in scale of 2-13%. In case of 0.5% N content in matter and the dose of 1ton of the digestate 5kg of nitrogen per hectare is applied into the soil. To use digestate as a fertilizer it is necessary to use only stable digestates produced by proper treatment practice. The most essential factor that has to be controlled is high content of organically bounded nitrogen. It is important to pay attention to initial C:N ratio of the feedstock (Mendelova zemědělská a lesnická univerzita v Brně, 2008).

Solid form characterization

To characterize amendment, fertilizing and innocuousness properties there are values of main indicators, like organic matter content, dry matter, C and N, P, K and other substances content, humic substances and impurities contents according existing standards or standards for similar materials.

Organic matter content – organic matter content varies according type of the feedstock. According Teglia et al., 2011 collected data from literature, organic matter content in digestate varies from 40-85% of DM, see Table 5. Organic matter as well as dry matter is important to determine amendment properties. According European standards on organic amendment properties the minimal organic matter content is 20% of DM for digested material. It shows that all digested materials can be considered as soil amendments materials, but each European country has own legislation, so mentioned values are only informative.

Table 5: Organic matter of different wastes (Teglia, et al., 2011)

Inputted waste	OM(% DM)
Dairy manure + biowaste	69–76
Organic fraction of MSW + pig slurry	68–71
Pig slurry + milk serum + cow slurry + maize silage + rice residues	70
Pig slurry + blood industry residues + maize silage	67–74
Primary sludge	55
Organic fraction of MSW	55
Mixture of primary sludge + organic fraction of MSW	58
Pharmaceutical industrial sludge	70
Cattle manure	86
Mixture of primary sludge + organic fraction of MSW	70
Food wastes +landscape wastes	39–43
Energetic crops +cow slurry+ agro-industrial waste + organic fraction of MSW	75

Nutrients content – content of carbon and nitrogen is very important mainly balance between organic and inorganic forms. C and N are improving biological properties of the soil. Total organic C content in investigated digestates is 400g/kg DM and N content is variable in scale of 50 - 150g/kg DM. Majority form of nitrogen presented in digestate samples is $N-HN_3$, always presented in value 50% and more of total nitrogen (Table 6), (Teglia, et al., 2011).

Table 6: Nitrogen content in different digestates (Teglia, et al., 2011)

Nature of input wastes	TOC(g/kgDM)	TKN(g/kgDM)	N-orga (g/kgDM)	N-NH$_3$ (%TKN)
Dairy manure + biowaste	–	50–60	19–27	52–62
Energetic crops + cow slurry + agro-industrial waste + organic fraction of MSW	404	65	32	51
Organic fraction of MSW + pig slurry	378–397	135–151	48–53	63–68
Pig slurry + milk serum + cow slurry + maize silage + rice residues	367–383	83–103	38–41	54–61
Pig slurry + blood industry residues + maize silage	387–421	85–92	31–34	61–67

French standards for fertilizing properties of organic matter determine that total N, K_2O and P_2O_5 must have minimum content in fresh matter 3% and total higher than 7%. German regulations order that nutrient content in DM must be higher than 0.5% N, 0.3% P and 0.5% K_2O.

Legislative aspects – for use as a fertilizer

The current situation for a rising number of biogas plants, is the quantity of produced biogas digestate, requires legislative establishments, for the protection of air conditions, management of fertilizers and others. According Czech legislature definition, digestate is: the residue remaining in digester tank after anaerobic digestion process where the biogas is produced (Mendelova zemědělská a lesnická univerzita v Brně, 2008). Worldwide there exist many regulations applying on anaerobic digestion divider according waste used as substrate, type of facilities and finally use of by-products. The digestate is subject of waste management facility, that has to be licensed and if substrate is form animal waste, it has to be treated under the specific conditions (Monnet, 2003).

German regulations are so called the principles of good execution of fertilizing. No liquid manure or raw material is allowed to be spread to soil in season of November 15th to January 15th; this is the same in other countries. It is because of possible leakage of nitrogen rich liquid from the soil surface, which is not permeable because of frost. Next regulation covers digestate utilizing is no more than 210 kg/ha total N is allowed to be spread on pasture, no more than 170 kg/ha on land used agriculturally (Deublein, et al., 2008). Recycling of organic wastes demands regulations to control proper use of these materials. For digestate use of these regulations is different, respecting some elaborating limits. If digested material do not

	Maximum nutrient load	Required storage capacity	Compulsory season for spreading
Austria	170 kg N/ha/year	6 months	28 Feb – 5 Oct
Denmark	170 kg N/ha /year (cattle) 140 kg N/ha/year (pig)	9 months	1 Feb – harvest
Italy	170 – 500 kg N/ha/year	90 – 180 days	1 Feb – 1 Dec
Sweden	170 kg N/ha/year (calculated from livestock units per ha)	6 – 10 months	1 Feb – 1 Dec
Northern Ireland	170 kg N/ha/year	4 months	1 Feb – 14 Oct
Germany	170 kg N/ha/year	6 month	1 Feb – 31 Oct Arable land 1 Feb – 14 Nov Grassland

Figure 11: Nitrogen standards in different countries (Lukehurst, et al., 2010)

comply standards, it cannot be used in the soil (Makádi, et al., 2012). Figure 11 shows nitrogen nutrient load regulations in different countries and in which season can be the fertilizer spread. pH and C/N ratio – data shows that digestate common value is ranged between 6.7 and 8.5 (Figure 12). C/N ratio has been measured in scale of 7, for digestion residues from energetic crops, cow slurry, agro-industrial waste and organic fraction of MSW, to 25 for food and green wastes. C/N ratio is changeable and in most literature is lower (<20).

pH, TS and VS content for the samples of feed, digestate from the batch trials (intermediate stage), and digestate from the reactors for the systems under study

Sample		PS	OF	PS:OF
Feed	pH	5.23	3.4	3.6
	[VS] (gl^{-1})	42.8	50.6	52.5
	[TS] (gl^{-1})	61	54.8	64.3
Intermediate stage	pH	7.16	4.49	6.11
	[VS] (gl^{-1})	34.1	19.8	22.6
	[TS] (gl^{-1})	58.1	24.4	34
Digestate	pH	7.4	7.2	7.1
	[VS] (gl^{-1})	19.5	13.5	12.5
	[TS] (gl^{-1})	35.4	24.4	21.4
	VS reduction %	54.4	73.3	76

Figure 12: pH, PS and TS (Gómez, et al., 2005)

2.2.4 Quality management

Quality is assessed in 3 aspects according to Monnet, 2003: biological, chemical and physical. Chemical quality is identified by the presence of heavy metals and others inorganic contaminants, persistent organic contaminants (herbicides, pesticides or antibiotics) and nutrients. There can be even others hazardous matters like pathogens, seeds or TSE (transmissible spongiform encephalopathy). And other impurities, physical, like plastic and rubber, metal, glass and ceramics, sand, stones and cellulosic matter, that is decreasing digestate quality and represents the risk of its utilizing (Monnet, 2003). The digestate contains all material, which was not decomposed and converted to biogas, consequently even contaminants that were present in feedstock remains in the digestate. It follows, well prepared feedstock will produce a good quality digestate and vice versa. Even if the quality of digestate is quite high and digestate is used for agricultural purposes, whole or separated, the value for the producer is relatively low. Because of the legislative regulations about spreading nitrogen rich materials the digestate cannot be used locally. This is to avoid over application. Another challenge lies in the application season and suitable market. This results in transportation, storage, spreading and operational costs. Finally the digestate production can be more costly then profitable. For this reason, new

Physical	Thermal
Thickening (Belt)	Drying (Rotary Drying)
Thickening (Centrifuge)	Drying (Belt drier)
Dewatering (Belt press)	Drying (J-Vap)
Dewatering (Centrifuge)	Drying (Solar)
Dewatering (Hydrocell)	Evaporation (scraped surface heat exchangers)
Dewatering (Bucher press)	Conversion (Incineration)
Dewatering (Electrokinetics)	Conversion (Gasification)
Purification (Ultrafiltration and Reverse Osmosis)	Conversion (Wet air oxidation)
	Conversion (Pyrolysis)

Biological	Chemical
Composting	Struvite precipitation
Reed Beds	Ammonia recovery (Stripping + Scrubbing)
Biological Oxidation	Ammonia recovery (Membrane Contactor)
Biofuel Production (Algae)	Ammonia recovery (Ion Exchange)
Biofuel Production (liquor as process water)	Acidification
Biofuel Production (hydrolysis of fibre to Bioethanol)	Alkaline Stabilisation
Microbial Fuel Cell	

Figure 13: The digestate enhancement techniques (Frischmann, 2012)

techniques for digestate enhancement are invented. The goal is to increase the value of digestate, to make new market products, to decrease dependence no land application, to reduce operational costs and ensure more secure and suitable outlets for digestate products (Frischmann, 2012).

Pre-Digestion Enhancement Techniques of digestate are divided to physical, thermal, chemical and biological, see Figure 13.

For example dewatering is separation of the solid and liquid part of digestate to achieve separated fiber content of 18% and more of dry solids and liquor part. After dewatering 80% of mass is removed into the liquid part. It is the first process of the digestate treatment. Digestate fiber is semi-solid, means easier to be stored (Frischmann, 2012).

The biological treatment of digestate sets its agricultural quality, which can be defined by 3 aspects:

1. Organic amendment properties: useful for restoring the quality of degraded soil. Amendment material is material which improves the physical properties of soil, like water retention, permeability, infiltration, aeration and structure. In general amendment material presents better environmental conditions for roots and for whole plant development.
2. Fertilizing effect: presents providing of both macro- and micronutrients to the soil, improving soil fertility and production yield of crop cultivation.
3. Innocuousness: related to issues of possible impact on environment or human and animal health (Teglia, et al., 2011).

Digestates can be directly returned to the soil as amendment or fertilizer, but in case of its instability, they need to be post-treated to improve quality. Quality standards already exist for composts, but there are new investigations to determine those for digestate and its direct agricultural use (Teglia, et al., 2011).

2.2.5 Other ways how to utilize digestate

Typically the digestate is used to farmland; next possible utilization is to use it like animal bedding, it means the separated part of digestate. Like a part of post-treatment procedure, is more frequently treated by aerobic process of composting (for use in horticultural applications). (Alexander, 2002) In present time is digestate tested for utilization as a solid fuel (Kratzeisen, et al., 2010). Thanks to research there will be surely new and new possibilities of digestate utilization and disposal.

2.3. Briquettes

Even the topic of work is not concerned to solid biofuels; the following paragraph is dedicated more or less to characteristics and properties of briquettes and pellets designated for energetic purposes. The processing of digestate is similar or evens the same to that used for biomass with energetic value. There are many different materials used for its energetic value, like a plant biomass, wooden biomass mixtures and others. These materials are processed into the form by piecemeal, crushing, pressing, and other ways. Before determining the size of material particles, moisture is adapted. Size (granulometry) is changed by crushing, cutting and sorting; and the moisture is decreased by drying to final moisture content 15-20%. Before burning of biomass biofuel; its form is changed by briquetting or pelleting technology. In comparison of these two forms, pellets have a number of advantages over briquetting. For example, pellets produce a stable and tough structure with a smooth surface; so it can be used automatically. In case of briquettes, its structure is crumbly and unstable, thus the mechanical application is excluded. In case of digestate agriculture use, there is not necessary to have all advantages of pellets, because it is supposed to be decomposed in the soil. Physical properties required for solid biofuels are specific weight, mechanical durability and the bulk density. Bulk density is defined as a mass of many particles of the material divided by the total volume and is gravimetric determined by free pouring of material into vessel of known proportions (VŠCHT, Ústav energetiky, 2012).

2.4. Sorption

2.4.1. Definition of sorption

Sorption can be described as a mass transport of gasses or liquids into solid material; it describes the penetration into the bulk phase of absorbing solid (Fletcher, 2008). Relevant sorption can occur in two ways WATER-SOLID or AIR-SOLID (Volz, 2009) and (Marešová, et al., 2012). Sorption can be divided to absorption or adsorption (Fletcher, 2008). The process in which a contaminant that adheres to a solid is adsorbate and sorbing a solid is adsorbent (Piwoni, et al., 1990). Adsorption is divided into physical and chemical categories, their characteristics are shown in Table 7. Physical adsorption, physisorption, is a dynamic process where molecules are held by dispersion forces or van der Waals forces. This sorption is dependent on the physical pattern of the adsorbent, like porosity; and occurs to varying extents for all adsorbates, increasing with decreased temperature and increased pressure. The process is always exothermic. Chemical adsorption, chemisorption, is less common and presents electrons transfer between phases with the formation of chemical bonds. The essence is a chemical reaction causing the adhesion of adsorbate molecules. Chemical bond regeneration is difficult and desorption (reversible process of adsorption), means that re-usage is impossible. This fact causes a yield of chemically different products against the initial one.

Table 7: Characteristics Associated with Physical/Chemical Adsorption (Fletcher, 2008)

	Physical Adsorption	Chemical Adsorption
Heat of Adsorption / kJmol-1	20 - 40 c.f. heats of liquefaction	> 80 c.f. bulk-phase chemical reactions
Rate of Adsorption (at 273K)	Fast	Slow
Temperature Dependence of Uptake (with Increasing T)	Decreases	Increases
Desorption	Easy- by reduced pressure or increased temperature	Difficult - high temperature required to break bonds
Desorbed Species	Adsorbate unchanged	May be different to original adsorptive
Specificity	Non-specific	Very Specific
Monolayer Coverage	Mono or multilayer condition dependent	Monolayer

2.4.2. Porosity

Most materials are to some extent porous. Physical properties, e.g. density, strength, conductivity of solids are dependent on pore structure (Figure 14) and its control. Industrial world is largely using the porous materials like catalysts, industrial adsorbents, membranes and ceramics. Pores present passages (from the Greek word 'πoροσ' - passage) between external and internal surfaces of solid. All materials

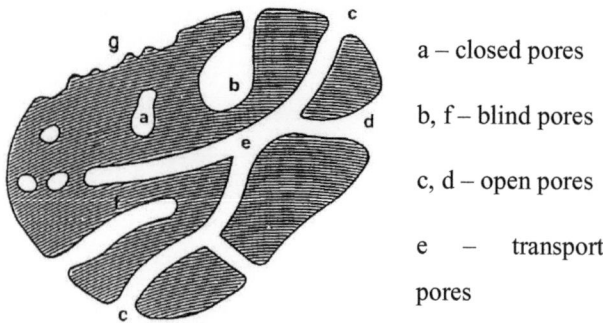

a – closed pores

b, f – blind pores

c, d – open pores

e   –   transport pores

Figure 14: Pores (Commission on colloid and surface chemistry, 1994)

possess pores are used to sorption purposes, such purification or separation. Surfaces of solid material are described as external and internal (Fletcher, 2008). An external rough surface is not porous, because it does not have irregularities deeper than it is wide. (Commission on colloid and surface chemistry, 1994). Sorption filling process can be described as 1) monolayer formation, 2) pores filling by co-operative effects, 3) completion of pores filling process (Fletcher, 2008). Mechanism of sorption includes chemical adsorption, multilayer adsorption, capillary and co-operative pores filling, phase transition, molecular ordering and heterogeneity of solid surface (Gruszkiewicz, et al., 2005).

Pores can be classified in four terms: open (connected to external surface of solid – passing for adsorbate), closed – are influencing macroscopic properties like bulk density, mechanical strength, but are inactive in flow processes (internal void, is not connected to external surface), transport (is connecting different parts of external surface to internal) and blind (connected to transport ones, do not lead to external),

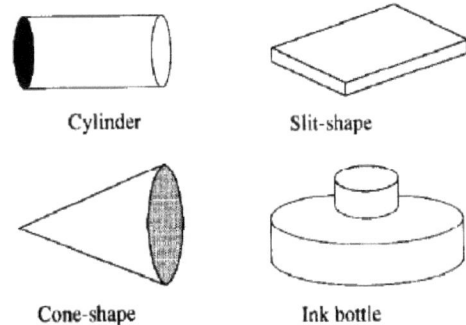

Cylinder   Slit-shape

Cone-shape   Ink bottle

Figure 15: Pores shapes (Kaneko, 1994)

see Figure 14. Pores can also be classified according to their shape and width and routes. In terms of shape, pores are cylindrical (e.g. in zeolites), ink-bottle, funnel or slit-shaped (Figure 15).

An idealized system describes shapes as cylinders, prisms, cavities, windows, slits, spheres (Commission on colloid and surface chemistry, 1994). But real porous systems are much more complicated. They consist of a variety of complex shapes, connections and distributions and different sizes of pores within one material. According to accepted IUPAC (International Union of Pure and Applied Chemistry) classifications the guideline of pore widths is:

o Micropores: < 2nm

    Ultramicropores: < 0.5 nm

    Micropores: 0.5-1.4 nm

    Supermicropores: 1.4-2.0 nm

o Mesopores: 2-50 nm
o Macropores: > 50 nm

Micropores are results of imperfect stacking of molecules and packing arrangement of bulk matter, looking like series of interconnecting volume elements, but with each element different size and shape. Micropores are providing maximum adsorption potential. Sorption in this case is completely reversible.

Mesopores contain defects in the solid structure and provide passages to ensure there is a transport system to micropores. They are filled in last stage of isotherm by multilayer formation. Determination of mesopores is via vapor adsorption.

Macropores serve as transport pores allowing access to internal surface. Macroporosity can be observed by optical microscope and scanning electron microscopy. Usually the diameter of the pores is 1-2mm (Fletcher, 2008).

Origins of pore structure

Not everyone adsorbent contains all types of pores, and pores are formed in different ways. Every porosity structure is unique and pores are not uniformly shaped (Fletcher, 2008). Some porous powered materials, so called agglomerates, are consolidated by many others macroscopic bodies. Others are unconsolidated, so called aggregates are weakly packed particles. The particles themselves could be significantly porous or nonporous; they are surrounded by interparticle and internal voids.

Porous material can be created by several ways and pores are the inherent features of crystalline structure, like Zeolites and clay minerals (Commission on colloid and surface chemistry, 1994). Zeolites are widely used adsorbents, known as aluminosilicates and are used in their artificial or natural form. The mechanism of water adsorption in zeolites is not clear, but it seems that water molecules are keeping a range of energies to variety locations (Gruszkiewicz, et al., 2005). Pores sizes in zeolites are usually less than 1 nm and so these adsorbents are used in drying processes of the air, CO_2 purring from natural gases and others (Pechoušek, 2010).

In general, the prediction of adsorption of any material is difficult, because of the lack of accurate characteristics of a pore system even for the most uniform sorbents (Gruszkiewicz, et al., 2005).

Quantitative description

Total pore volume of solids (V_p) is volume of pores, measured by defined method, which is stated.

Pore size (or width) is distance between two opposite walls of the void – in micro-, meso- and macro- scale.

<u>Distribution of pores</u> size is represented by derivates $\frac{dA_p}{dr_p}$ or $\frac{dV_p}{dr_p}$ as a function of r_p – A_p wall area, V_p volume and r_p radius of pores.

<u>Porosity</u> (ε) is ratio of total pore volume to apparent volume of particle in both open and closed pores. (Commission on colloid and surface chemistry, 1994)

For porosity measurements and determinations there are some methods how to investigate pore sizes and distribution, these methods are volumetric, gravimetric, calorimetric or spectroscopic. Sorption is recorded as graphically, known as isotherms (Pechoušek, 2010).

Soil sorption

In the case of agriculture, they utilize mainly the sorption system WATER-SOLID which means, the ability of the soil to attract ions and whole molecules of water or solutions into solid form of the soil (Richter, 2004). Water molecule has polar charge, and is very well arranged to the mineral surface; because of the negative charge on most of surface. Water in the soil is presented as adsorption water (hygroscopic), viscous water (capillary) and free water (ROADEX, 2011). These catch solutions with nutrient content and make a reservoir of nutrients during the vegetation season. Every soil has a different method of sorption of nutrients: mechanical, physical, chemical, physic-chemical and biological (Richter, 2004).

3. OBJECTIVES

Based on the knowledge of the process of digestion and its products, from the available literature sources were created objectives of this work. The aim is to describe the basic characteristics and results of mechanical and physical sorption of water produced from the biomass under different external conditions and thus get an idea of the potential use of compressed organic material. The factual content of the book represents three sub-goals.

The first of the goals was to increase awareness of basic, mainly mechanical, properties of the compressed digestate both information from secondary sources, and laboratory analysis of samples collected from two biogas plants in Czech Republic. These properties are important for assessing the suitability of digestion products, after subsequent mechanical treatment, as appropriate fertilizer in the form of briquettes. There is also the possibility to consider using additional materials, such as dolomite limestone or zeolite, and thus improving the current fertilizing properties and distribution of nutrients to the environs. Alternatively, offers the possibility of use in heavy soils to improve the physical structure of the soil. Properties of digestate briquettes will be compared with other materials.

Another objective is to determine by laboratory experiments sorption potential of the digestate briquettes, again in comparison with a different material. Knowledge of sorption is important because of the possibility of changing mechanical properties of briquettes after incorporation into the soil. At present, the digestate is used traditionally in the form of liquid fertilizer, in the form, which leaving the biogas plant and in order to achieve negligible losses of nitrogen is immediately incorporated into the soil. However, the liquid form must be stored in special conditions, which compressed form does not require.

Based on gained properties and measured values it is better to understand the mechanism of water uptake by briquettes under different conditions, and it is actually the third sub-objective, which is the most important part of the overall objectives of the book - changes in digestate briquettes properties depended on water uptake. Along with the acquired knowledge of digestate in the form of briquette, may be better to propose other use of digestate such as organic fertilizer or as a means to modify the physical properties of the soil. In this study, the economic performance and profitability of innovative digestate briquettes using will not be taken into consideration. However, it could be further explored in future research on use biologically degradable materials such as digestate, the final substrate after the process of anaerobic digestion.

This research is entirely focused exclusively on the mechanical properties and sorption of granulated organic waste, which has not yet been observed.

4. MATERIALS AND METHODS

4.1 Used material characteristics

During the experiment, following materials were used: pure digestate, hemp, sorghum, spruce bark and various mixtures of materials like sorghum- spruce bark, sorghum-shavings, then digestate mixtures with additives such as dolomite limestone and zeolite. The most important material for the experiment is digestate in form of briquettes. Other materials have been used mainly due to compare their properties with the digestate and they were the most available within the laboratory.

Used digestate derives from two different commercial biogas plants, collected in autumn of the year 2012, ZEPOS a.s. Radovesice (D1), Czech Republic and agricultural cooperative Krásná Hora nad Vltavou, the farm Petrovice (D2), Czech Republic. The composition of the feedstock used in process of digestion within biogas plant (sample D1) is: 60% corn silage, 10% of pig slurry and 30% manure. Feedstock for sample D2 is - 20% corn silage, 20% grass silage and 60% beef manure from the farm. Both biogas plants are large-scale, with an output of 800 kW, using technology of company FARMTEC a.s.

Sample D1 was separated, before analysis, by a sieve to liquid and solid form. The separated solid part was re-dried in the laboratory at 100% of dry content.

Sample D2 was already in the solid form, partially dehydrated with 78-80% moisture content. Both samples were analyzed in the laboratory of Czech University of Life Sciences Prague (CULS). The analysis results are presented in table form in the first chapter of Results.

4.2 Briquetting process - treatment of raw material

Before pressing procedure, all materials are, except the digestate which solid particles go through the eye of sieves, pulverized by a hammer crusher to lower parts with diameter of 8 mm. All of these materials have undergone a process of granulation using a piston press with a diameter of 62 mm, which is available at Technical Faculty of CULS.

Storing of all materials, before and after processing, was carried out at ambient room temperature. Before the material is compressed, it must be dried. Drying is then carried out, under ambient conditions, and then by a dryer at temperatures up to 95 ° C. The final moisture content of the material is equal to value of 14-15%. Further material can be compacted and granulated (briquetted) into the final shape of a cylinder with dimensions: diameter D = 6.2 cm, and different lengths (L). Briquettes have laboratory humidity 8-10%, and are prepared to test their mechanical properties. To make briquettes held fixed shape, they go through a series of tests, one of which is testing of an abrasion. The percentage of crumble briquette material is measured, directly dependent on time and the repeating of the test. Abrasion is measured by the crumble drum into which are inserted a pre-weighed samples - let the crumbling for 5 min, the difference in weight after the finishing the crumbling defines the cohesion of briquettes.

4.3 Experiment methods

4.3.1 Physical (mechanical) properties

To describe the physical and mechanical properties of digestate, samples were analyzed in the laboratory of CULS - and values obtained were tabulated.

- Relate to the determination of the following quantities is given their the percentage of the ash, nitrogen, fats, fibers, organic matter, Nitrogen Free Extracts (NFE) and gross calorific value (the amount of heat released by the complete combustion of fuel in a pressure vessel in the calorimeter, in environment of compressed oxygen at a temperature of 25 ° C - ČSN ISO 1928)

Other physical and mechanical properties of briquettes digestate were measured or calculated, and these are:

- Density: using the formula for the calculation of density (Eq. 2) and measured values of mass (measured in Figure 16) and volume (Eq. 1), density is calculated. Values of densities of all material briquettes samples are tabulated and compared.

$$V = \frac{\pi D^2}{4} \times L \ [m^3] \qquad (1)$$

$$\rho = \frac{M}{V} \ [kg/m^3] \qquad (2)$$

Figure 16: Weighing briquettes

- Bulk density (BD): into a container of known proportions (Figure 17) is poured uncompressed digestate and aligned simultaneously to the line with the container. Then its weight is gravimetrically determined and form measurements of the weight, container size, thus computed volume, the bulk density (Eq. 3) of uncompressed digestate is calculated.

Figure 17: Bulk density measuring

$$\rho_S = \frac{M}{V} \ [kg/m^3] \qquad (3)$$

- Moisture content - is measured using a standard laboratory hygrometer or calculated from the gravimetric method.

4.4 Monitoring of water uptake briquettes in the laboratory

Sorption experiment was carried out in the laboratories of CULS, room № C63, in the period November-April 2012/2013. An experiment is composed of three parts.

4.4.1 Sorption without limiting environment

Equipment: briquettes made of various materials. In this part of experiment were used these materials: pure digestate, hemp, sorghum, spruce bark and mixtures of materials like sorghum-pine bark, sorghum-shavings, then digestate mixtures with additives such as dolomite limestone and zeolite. Other tools are wire baskets with proportions app. 10x11cm (Figure 18), tray into

Figure 18: Equipment

which water is poured, and serves as a source of moisture. Stopwatches used for sorption time measuring. Laboratory scales, a common meter or calipers and prepared tables for data noting.

Practice: water is poured into the tray with app. 1 cm level. Before putting briquette sample into the water, the initial values are measured; diameter, length and mass (D, L, M). Sample for better handling is put into the water in a wire basket; allowed for three minutes, briquette sorbs water by its base. After three minutes, sample is removed from water and measured again. Measurement of D and L is always repeated five times. These five values at the end make a geometric mean (Eq. 4). Time of manipulation within measurement is 3 minutes, thus total period time is 6 minutes. This procedure is repeated until the samples are saturated with water. The point of saturation was conducted to change of increase in mass is less than 5%. All measured values are recorded in prepared tables for easier processing of data. The resulting graphs describing advancing sorption are listed in the Results chapter. Used calculations are - the geometric mean for repeated measurement of diameter and length, for data processing was used text editor MS Excel, statistical functions in it GEOMEAN (geometric average), see equation:

$$G(x_1, x_2, \ldots, x_n) = \sqrt[n]{x_1 \cdot x_2 \cdots x_n} = \left(\prod_{i=1}^{n} x_i \right)^{\frac{1}{n}} \qquad (4)$$

4.4.2 Sorption in partially limited environment - limited diameter

The second part of experiment was carried out in laboratory conditions.

Equipment: compared briquettes of digestate (n=3) and hemp (n=3) of similar mass, to compare possible differences in sorption. Furthermore, two polyvinyl-chloride (PVC) tubes (Figure 19) were used, with proportions:

$D_1 \times L_1 = 7.7$ cm x 17.7 cm

$M_1 = 342$ g

$D_2 \times L_2 = 7.7$ cm x 17.50 cm

Figure 19: PVC tubes

$M_2 = 341$ g

Furthermore equipment needed: tray for water, meter, scales, and stopwatches to measure the time of sorption.

Practice: first, initial values of briquette must be measured; length, mass and diameter – repeating of one value measurement by seven times, which is calculated as the geometric mean. Briquette sample is placed into the PVC tube, simulating the limited environment, in diameter level. The set briquette-tube is immersed into the water (about 2 cm level) and allowed adsorb water for 15 minutes. After 15 minutes, set is removed from the water and measured; the length of briquette is dedicated reading the Equation 5. Diameter is not measured because it restricted diameter tube, with the final value of 7.7 cm. Mass is measured in consideration of the PVC tube (Figures 20 and 21), net mass of briquette is given by reducing the tube mass. After measurement of all the required values, the procedure is repeated with the exception of the duration, periods are 30 minutes, 1 hour, 2 hours and 4 hours. The total time of the experiment takes app. 8 hours. All values are reported to the table graphically designed, presented in the Results chapter.

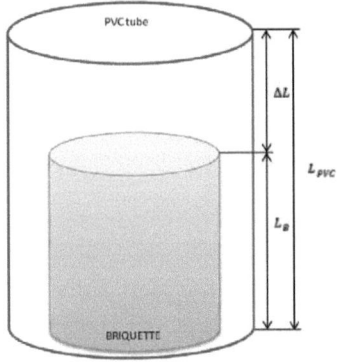

Figure 21: PVC tube complex with briquette

Figure 20: Measuring the length

$$L_B = L_{PVC} - \Delta L \quad [cm] \qquad (5)$$

4.4.3 Sorption in completely limited environment - simulation of soil conditions

The third part of the experiment was carried out again in the laboratories of the CULS. The entire experiment took place on 3rd - 11th April 2013; the number of measured days was 5-6.

Equipment: two wooden boxes with parameters of 26 cm x 56.4 cm x 26 cm, plastic foil, fine white sand, soil, humidity meter. Compared material is digestate and hemp briquettes.

Procedure: to the bottom of wooden box plastic foil is laid, like protection against getting wet the wooden material. Then pour to the middle row of white sand. On the sand the row of briquettes is longitudinally align; see Figure 24. Briquettes are packed with a soft absorbent paper for better water uptake and backfilled with white sand, for better recognition of enlarging mass, shown in Figure 23.

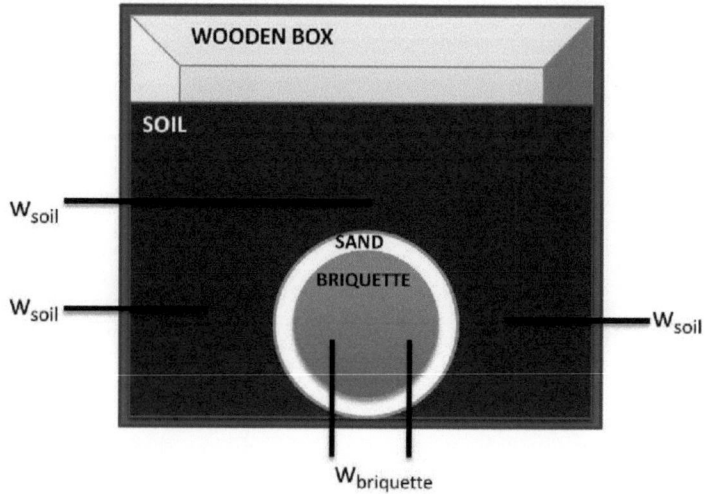

Figure 22: Wooden box with briquette, moisture measuring

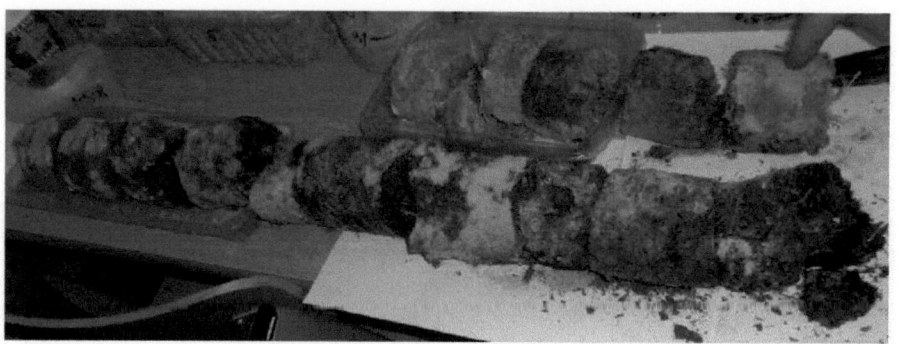

Figure 23: Increase in proportions

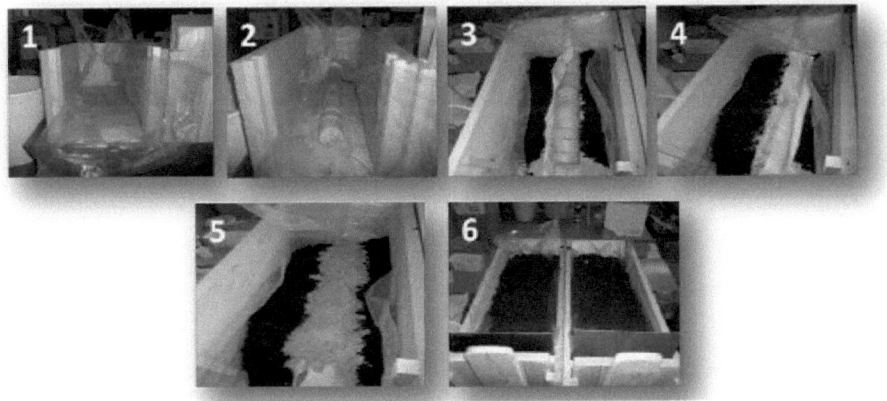

Figure 24: Wooden boxes and moisture observation, soil conditions

Briquettes are gradually covered by soil, about 25-30 cm. Soil is then watered sufficiently. In next days, gradually the layers of the soil are sliced along with briquettes. The humidity is measured (Figure 22) by humidity meter inside briquette ($w_{briquette}$) and external environment (w_{soil}), increase in briquettes diameter is measured as well. Changes are photo-documented. The values were recorded for subsequent evaluation and graphical processing - specified in the Results chapter.

4.5 Mathematical calculations

Used calculation for expression of total sorption (Eq. 6) of different materials:

$$V_{sorp} = \tilde{X}M_{tn} - \tilde{X}M_{t0} \qquad (6)$$

χ – Median t_0 – initial time

V – Volume t_n – final time

M – Mass

Determination of the maximum relative error of measurement – δ [%]: Maximum relative error of the measurement result is in the range from 2% to nearly 12%, depending on whether the corresponding value is measured directly or calculated from other measured quantities. Furthermore, the example of calculation errors of density documents the process and the resulting error.

For example, the average values of the measured briquette samples of different materials are:

D = 65 mm – diameter measured with an accuracy of 2 mm

L = 52 mm – length measured with an accuracy of 3 mm

M= 150 g – mass measured with an accuracy of 0.2 g

Calculations for density and volume:

$$V = \frac{\pi D^2}{4} \times L \ [m^3] \qquad (7)$$

$$\rho = \frac{M}{V} \ [kg/m^3] \qquad (8)$$

V – Volume   M – Mass

D – Diameter   ρ - Density

L – Length

The equation for calculating the measurement error: see for example the address below:

http://fyzika.upol.cz/cs/system/files/download/vujtek/texty/pext2-nejistoty.pdf

$$\delta_V = 2 \times \delta_D + \delta_L$$

$$\delta_M = \frac{\Delta M}{M}$$

$$\delta_\rho = \delta_M + \delta_V$$

Calculation of the relative error:

$$\delta_D = \frac{2}{65} = 0.03 \times 100 = 3\%$$

$$\delta_L = \frac{3}{52} = 0.057 \times 100 = 5.7\%$$

$$\delta_V = 2 \times 0.03 + 0.057 = 0.117 \times 100 = 11.7\%$$

$$\delta_M = \frac{0.2}{150} = 0.0013$$

$$\delta_\rho = 0.0013 + 0.117 = 0.118 \times 100 = 11.8\%$$

This is the resulting <u>maximum</u> relative error in the determination of density. Under specified conditions. This does not mean that such an error will be calculated for each density value.

5. RESULTS AND DISCUSSIONS

5.1 Analysis of digestate briquettes

5.1.1 Nutrient composition of digestate matter

On basis of the digestate matter analysis, the tables of observed nutrient values were compiled. The values were related to 100% and 6.2% of dry matter (DM) content. The values are expressed as percentage, see Tables 8 and 9. The table consists of two different types of digestate: D1, where the composition of digested feedstock is maize silage (predominantly), and cow manure and pig slurry. The digestate D2 is composed from maize silage, grass silage and the most cow slurry. The abbreviations in the Tables 8 and 9, meaning: NFE – Nitrogen Free Extract, OM – Organic Matter, GCV – Gross Calorific Value. GCV is related to 8-10% moisture content of the briquettes.

Table 8: Nutrient values of digestate in 100% DM, expressed in %

Sample	Ash	Nitrogen	NFE	Fats	Fiber	OM	GCV [MJ/kg]
D1 – lq.	30.65	19.98	40.23	1.47	7.66	69.35	17.32
D1 – s.	25.10	14.23	36.92	0.94	22.80	74.90	17.52
D2 – s.	17.12	13.08	38.23	0.38	31.18	82.88	18.55

Table 9: Nutrient values of digestate in 6.2% DM, expressed in %

Sample	Ash	Nitrogen	NFE	Fats	Fiber	OM
D1 – lq.	1.90	1.24	2.49	0.09	0.48	4.29
D1 – s.	1.55	0.88	2.28	0.06	1.41	4.64
D2 – s.	1.07	0.81	2.37	0.02	1.93	5.14

The liquid part of D1, the 100% DM digestate, contains 30% of ashes, 20% of useful nitrogen, and more than 60% of organic matter. These values are important to express mineral and organic significance of the organic matter, which is used in agriculture. The nutrients values of solid part, of D1, are not significantly different, compared to the liquid. Only the values of mineral nutrients are lower and values of fiber and organic matter content are higher. Differences are in range of 5-15%. The GCV represents the heat available from the fuel, which is completely combusted, expressed in MJ/kg (CHP Focus, 2013). The Gross Calorific Value of the D1 and D2 digestate is varying around 16-18MJ/kg. The D2 digestate is presented only in solid form (partially separated and dried to 75-80% DM). The nutrients values are not significantly different, varying in range of 1-9%.

In 6.2% DM, the representation of nutrients is low, in comparison to high dry matter content. The main component of this liquid matter is organic material with 4-5%, then Nitrogen Free Extracts with 3%. Samples D1 and D2 are similar in nutrients composition.

Result:

The composition of nutrients in samples of digestate is dependent on anaerobic digestion feedstock character, represented by differences between two special samples of digestate. Representation of important nutrients is higher in the 100% of dry matter, in order of 30%.

Discussion:

In comparison of organic matter contents with other results of the research, we can compare similar or totally different materials, in connection to initial AD feedstock composition. Measured organic matter in research of Paavola, Rintala (2008), is varying by 69-76% DM, where the investigated AD feedstock was pig slurry. Another collective of authors collected the values of organic matter from different initial feedstock. Final measured values are 70% DM in case of use pig slurry, milk serum, cow slurry and rise residues. I case of use just cattle manure the value was 86%. In summary of all the materials used for anaerobic digestion the range of organic matter content is between 40-85% (Teglia, et al., 2011). Nitrogen content in this paper is around the 20% in 100% DM. In study of digestates (Alburquerque, et al., 2012), the author sets the nitrogen level in cattle manure, which has higher dry matter content, in com parison to digestate, with very low dry matter content. Higher value of nitrogen in matter has the cattle manure with difference of 2%. In (Teglia, et al., 2011) study the content of nitrogen measured for different materials was 5-15%. In this paper the N content is between 14-19%. Abubaker, et al. (2012) compared fertilizing properties by observing nitrogen and other nutrients values. In case of comparison of NPK fertilizer with pig slurry digestate, the solid NPK had definitely higher value of 30.3%, besides the digestate had only 0.53%.

Calorific value of used samples of digestates in this study, is comparable with those used in Kratzeisen, et al., (2010) study, where used digestate was from maize silage, grass silage, potatoes and maize silage, sudan grass silage, poutry manure and corn mix. Investigated Calorific values were 17.3 MJ/kg and 16.4 MJ/kg, with 9% of moisture. Ash content was 18.3% and 14.6% within the 80-85% DM. In comparicon to another study, where the hemp was main material, the GCV was also 18 MJ/kg (Mankowski, et al., 2008).

5.1.2 Physical and mechanical properties of digestate briquettes

The first observed property of digestate briquettes is density. The density of digestate was compared with another material, for better imagination of its structure. All materials density is represented in Table 10. The average value of group of samples is expressed by statistical function, the geometric mean (Geomean), which is the most accurate for these measurements. Table 11 refers about digestate bulk density. Density is influenced by mass and volume changes.

Table 10: Bulk density of digestate uncompressed

Material	ρ_s [kg/m^3]	Geomean [kg/m^3]
Digestate bulk density	123.42	123.42

Table 11: Comparison of densities of different materials,

Material	ρ [kg/m^3]	Geomean [kg/m^3]
Pure digestate briquette	640.97 – 945.52	**814.06**
Digestate - CaCO$_3$ mix. b.	1,159.60 – 1,323.78	1,245.35
Digestate - Zeolite mix. b.	1,049.91 – 1,116.82	1,082.84
Miscanthus sinesis b.	726.95 – 816.03	783.25
Sorghum-pure	-	709.46
Sorghum + pine chips	-	885.16
Hemp	-	946.71
Spruce bark	-	1,034.21

The density in the digestate used in this book varies around 814.06 kg/m^3, the bulk density of non-briquette form is 6.6 times lower. The highest density within the Table has briquette of digestate mixed with crystalline structures, like dolomite lime or zeolite.

Discussion:

Dissimilar materials have different densities according to processing of briquetting. Thus the density varies by the changes of physical structure of the material. The concentration of individual internal components is changing as well. Chemical structure is not changing. When we compare the digestate briquette density with water density in the same ambient conditions, the density of water is higher by 186 kg/m^3, the water density is 1,000 kg/m^3 (4°C, normal pressure), (VŠCHT, 2008).

Zeolites have crystalline structures with high internal surface area characterized by a framework structure that encloses interconnected cavities (READE ADVANCED MATERIALS, 1997). Zeolites are very effective material, used for its excellent physical-chemical properties – high porosity and ion changes Zeolites have large internal surface area of up to 1000 m^2/g and volumetric density of about 0.8 kg/dm^3, (Maier-Laxhuber, 1991). By adding the zeolites into the soil, the soils properties will be changed. This material positively affects levels of trace and biogenic elements in the soil (BIOCLEAN, 2013). Density of the zeolite is given by its internal structure, the free internal space is smaller, and it will influence the sorption of water. The dolomite lime stone density is 250 kg/dm^3. (Thomas, 2013)

Initial moisture content is next observed value. The initial moisture of all used briquettes was 8-10%, and is influenced by around conditions of laboratory. Briquettes used for agricultural purposes, not to fuel purposes, can be storage in internal ambient conditions as well in external conditions with protection against direct exposure to precipitations. In case of exposure to water contact, briquettes lost its compatibility and shape stability.

Conclusion:

Thanks to density, we can briefly describe the mechanical properties of different materials briquettes, and gain the image of possible sorption. Density is dependent on force of granulating and the composition of material used as substrate for granulation. With the change of density, the mechanical properties will change as well.

5.2 Sorption processes

For assessment of the sorption processes, the graphs with the dependency of measurable values (M-mass, D-diameter, and L-length) on time were created. For explaining the basic mechanical tendencies of sorption water to solids, were investigated three different experiments. In experiments were used different materials and different limitation conditions, the changes in time were transformed to the graphical form. The accuracy of measured data includes the maximal error around 10%.

5.2.1 Sorption in non-limited conditions

In this part of the experiment were measured values changed during the time period in laboratory conditions, without any limiting factor for the sorption. Water has advanced to the briquette via the base. The progress of the sorption is graphically described in Figure 25. Describing values are volume (V_{sorp}) of sorption expressed as the difference between the initial mass of briquette (M_{t0}) and the last measured mass (M_{tn}) of fully sorbed briquette, see Equation 9. Total volume of sorbed water is expressed. Saturation point of briquette is stated in this paper like change of volume is less than 1% in the time period.

$$V_{sorp} = \tilde{X} M_{tn} - \tilde{X} M_{t0} \qquad (9)$$

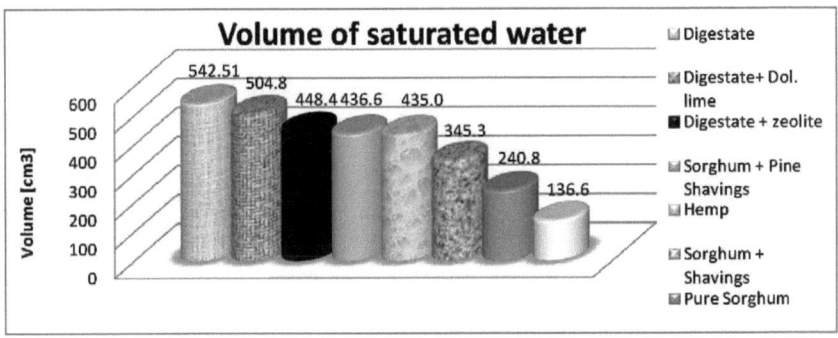

Figure 25: Total volume of water sorbed by different type of material

From the Figure 25, we can see the differences between volumes of each briquette according to used material, where the most saturated briquettes were those of digestate. The volume of saturated water in digestate briquettes was 542.5 cm3. The digestate had the smallest initial weight. The initial moisture of all material briquettes is the same, 8-10%. The last phase of the sorption is characterized by less adhesive forces of the briquette structure, pieces of matter are ripping caused by manipulation during the measurement.

Table 12: Initial mass of briquettes (average values of initial mass)

Digestate	Spruce bark	Sorghum	D+Zeolite	Hemp	Sorg.+pine chips	D+lime	Sorg.+ch ips
124.4 g	126.3 g	141.9 g	149.3 g	152.0 g	165.5 g	169.0 g	170.3 g

Initial mass of briquettes is recorded in Table 12, value of mass is different, but by this way of investigations of the sorption, is not possible to determine, if the change of sorption character is dependent on the initial weight of the briquette.

In Table 13 are counted values of last saturation time for all the materials (see Table in Annexes). Median represents the middle time value of all samples, when the saturation was ended. Modus is the most frequent time value. Minimum time of the end saturation is 21 minutes and maximum 165 minutes. In case of all 5 samples of digestate, the end saturation time value was in average 141 minutes. And the time ranged between 105th – 165th minute.

Table 13: Saturation times

Statistical function	Median	Modus	Geomean	max	min
Time [minutes]	105	105	88	165	21

Conclusion:

The ability of sorption is given by mechanical and physical properties of each material. These properties are influenced by the type, size, and shape of elements inside the briquette. Each granulated material can be considered as unique sample, which is not homogeneous. From the figure the volume of water sorbed by each material is comparable. The digestate in this experiment has proved as the most sorption able medium. Digestates with additives (zeolites and the lime) have smaller volume of sorbed water. It can be explained theoretically as result of high density. The higher density, the smaller internal space, caused fine particles of crystalline structure.

There is no other research concerned to description of the sorption properties and mechanisms of such kind of material, like the digestate. For determination of the sorption potential and its dependency on the initial mass of briquette is needed. The statistically confirmed results can be presented as the certified sorption principles.

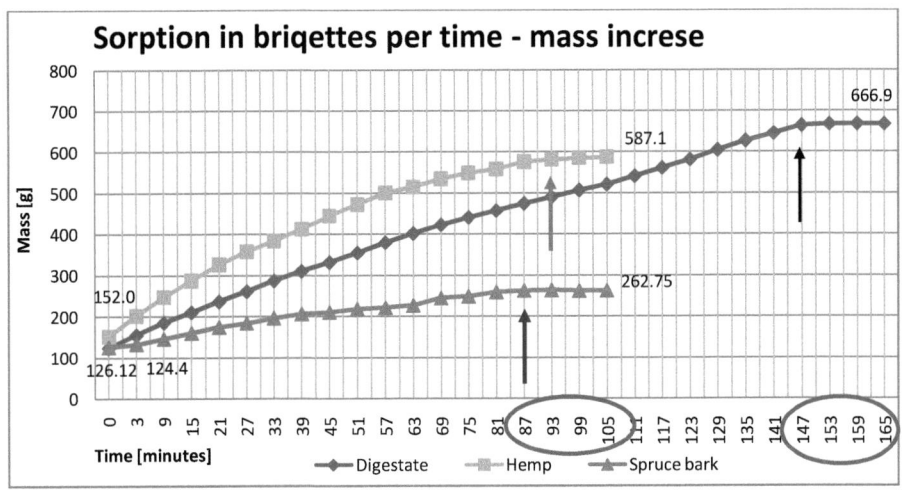

Figure 26: The time course of sorption for digestate, hemp and spruce bark

Figure 26 records the timing and mass, thus volume, increasing with advancing sorption. Compared materials were digestate, hemp and spruce bark.

The sorption of the digestate has longer time in comparison to both hemp and bark material. The saturation of the digestate briquettes in average starts in time of 147th minute, the slope of the graph started to be less than 1%. For hemp the saturation starts in 93rd minute and for spruce bark in 87th minute.

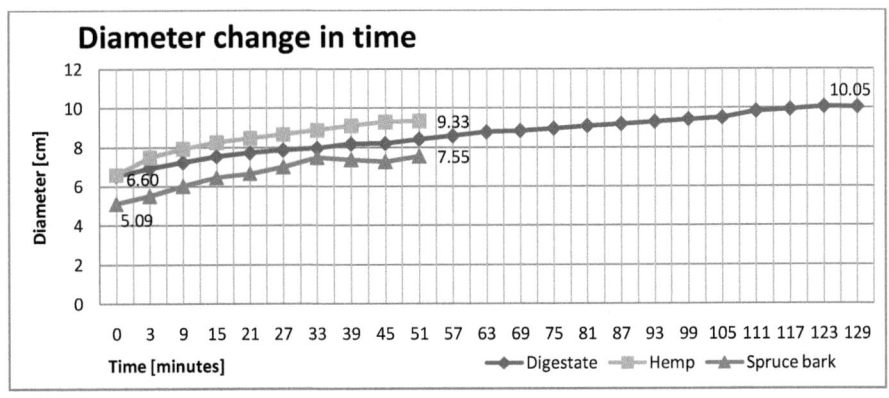

Figure 27: diameter changes in advancing sorption

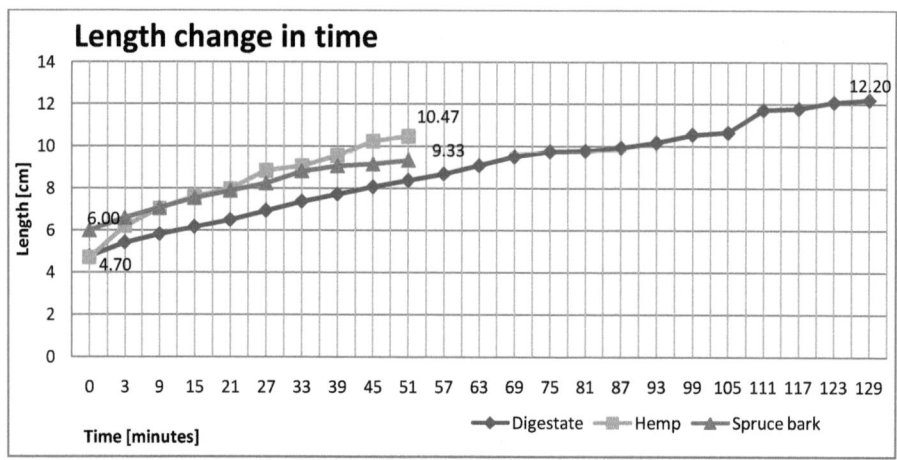

Figure 28: Length changes with advancing sorption

Upper Figures 27, 28 are describing the increase of diameter and the length. The digestate diameter increased from initial value of 6.6 cm to 10.05 cm, the difference is 3.45 cm. The diameter of hemp has increased by 2.73 cm, from the value of 6.6 cm to 9.33 cm. And the diameter of spruce bark has increased by 2.46 cm, from initial diameter of 5.09 cm to 7.55cm.

Next Figure is describing the differences in length. The initial length of digestate was 4.76 cm, which rose to 12.20 cm. The difference is 7.44 cm. Hemp briquettes length has prolonged from 4.70 cm to 10.47 cm, by the difference of 5.77 cm. And the spruce bark has prolonged by 3.33 cm, from 6.00 cm to 9.33 cm.

Conclusion:

Sorption process has been recorded to the graph as a raising curve in case of all investigated materials. The growth of briquette varies according to material, from 3.33-7.44 cm in length and 2.46-3.45 cm in diameter. So the average increase in proportions is 2.88 cm x 5.5 cm (DxL).

Sorption is longer in case of digestate, difference in time of saturation between digestate and hemp with spruce bark is app. 1 hour. The time of saturation between spruce bark and hemp is only 6 minutes. During those 165 minutes the briquette can gain up to 5 multiple of its initial mass. At the spruce bark briquettes the mass raised two times, in time of 105 minutes. And in case of hemp, mass rose 3.8 times, in 105 minutes as well.

The conclusion is that the digestate briquette exposure to sorption is increasing in all its proportions relatively quickly, in 165 minutes, in assuming permanent supply of water and non-limited conditions. The sorption can be hypothetically explained by the non-limited conditions, allowing the water to adsorb to internal space of the briquette. Sorption has resulted as an obvious and measurable change of proportions and shapes of briquettes. Mass, thus even density was influenced. With the change of basic physical properties, all properties of briquette are changed. The saturation varies according to materials used, characterized by density of material.

These results are mainly useful to illustrate the tendency of the sorption and confirmation of this tendency is possible on the basis of next similar research. The comparison of these results to other author's findings is missing because similar research has not been done.

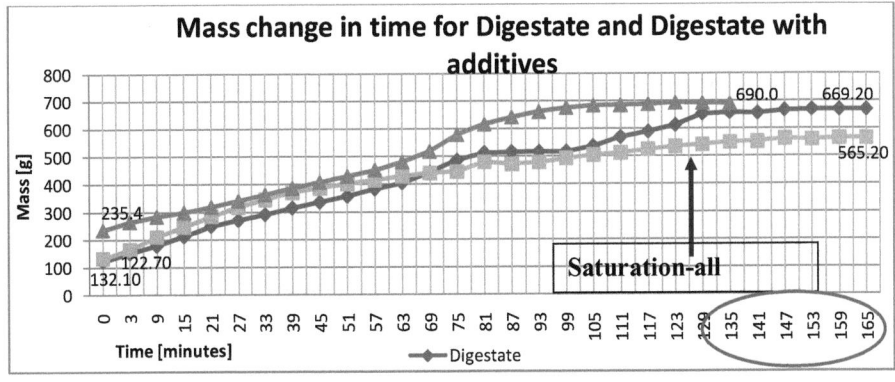

Figure 29: Comparison of sorption of pure digestate with additives

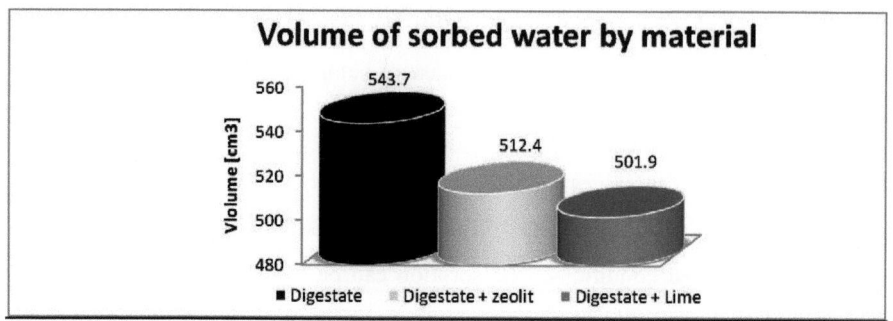

Figure 30: Volume sorbed

In comparison of two other materials, which are the mix of digestate with crystalline structure of zeolite and dolomite lime, we can observe how the sorption is influenced by internal structure of the material. Zeolites are known as industrial sorbents able to absorb the water relatively in short time. According to Ramos et al. (2003) the maximal water adsorption capacity of spherical zeolite pellets with diameter of 4 mm, is 300g of adsorbate to 1kg of adsorbent. The sorption is different in the beginning phase between the zeolite mixture and pure digestate, but after an hour the sorption curve of pure digestate still quietly raise and the zeolite curve stabilizes.

In comparison to digestate the difference is relatively obvious from the graph. The saturation, with the curve increase less of 1%, occurs within the all three materials around the time of 135th minute (Figure 29). The time interval for mixture of digestate with dolomite lime is 105-135 minutes. Next two materials, the pure digestate and mixture with zeolites, sorb water up to 165th minute. The volume of absorbed water is by 30-40 ml higher in digestate, comparatively to other two materials (Figure 30).

Conclusion:

Sorption is given by internal structure of the sorbent material which is with differently speed filled by water molecules. In this case, of non-limited conditions, digestate briquettes have sorbed more water than two other materials, in longer time of sorption.

5.2.2 Partially limited sorption – in diameter

The second part of the experiment is different from the first by using PVC tubes, into which is inserted a briquette, that is limited to the diameter growth of briquettes. This part of the experiment is monitored mainly by growth in length and mass change briquettes (Figure 31 and 32). Two materials compared - digestate and hemp.

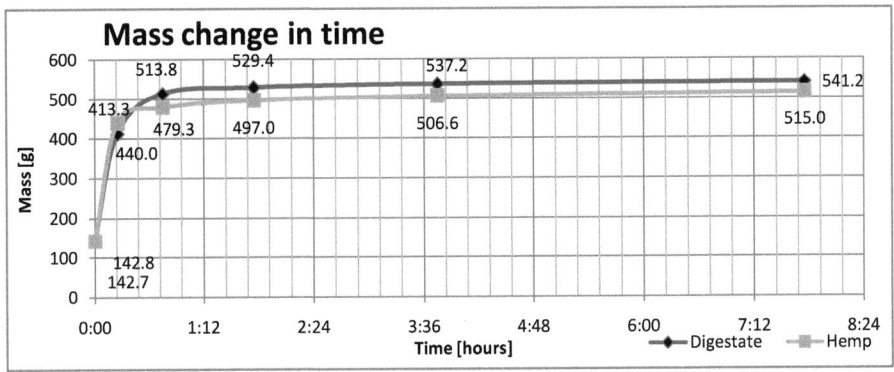

Figure 31: Mass change in time, digestate and hemp

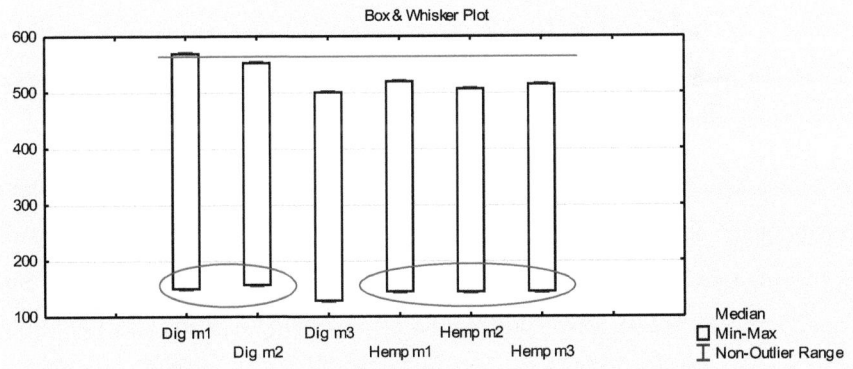

Figure 32: Comparison of mass increase excludes the time

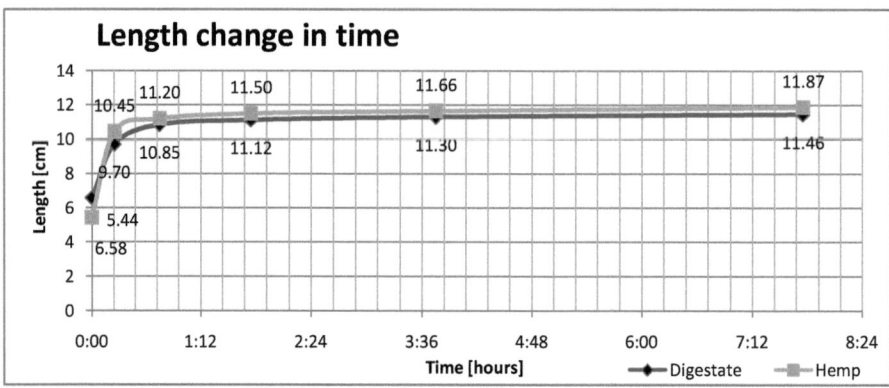

Figure 33: length change in time

At the beginning of sorption the digestate briquette weighed 142g, after the sorption within 8 hours the mass has increased to value of 541g. In measurement of hemp the initial mass was 142g as well and the final 515g, with the same time of sorption. The volume of sorbed water is 399ml for digestate and 373ml for hemp. The average difference in volume between hemp and digestate samples is in range of 30ml.The highest curve elevation was recorded in time of first 15 minutes, with the increase in mass of 70-80% in both cases. Then the slope of curve is decreasing to 45th minute by: in case of digestate 25%, in case of hemp 10%. After 45th minute the increases in curve slope are less than 4%. The saturation in digestate case occurs after 3 hours of sorption. In case of hemp it would be a little bit longer time, but the measuring has been stopped, when the slope of curve had more than 2% elevation.

Observed length has increased in average by 5.36cm at digestate and by 4.97cm at hemp briquette from initial to final value, Figure 32. The highest prolonging has occurred after first 15 minutes of sorption, in both material cases (digestate – 3cm, hemp – 5cm). Figure 33 shows comparison of prolonging of briquettes from initial almost same conditions excluding time, upper line indicates more sorption material.

5.2.3 Sorption in completely limited environment - simulation of soil conditions

The third part describes changes in moisture, volume and density of briquettes placed in the soil, thus in conditions limiting the space around line of briquettes and therefore its expansion. Advancing moisture of briquettes in comparison to soil moisture is recorded in relation to time in the following chart (Figure 34).

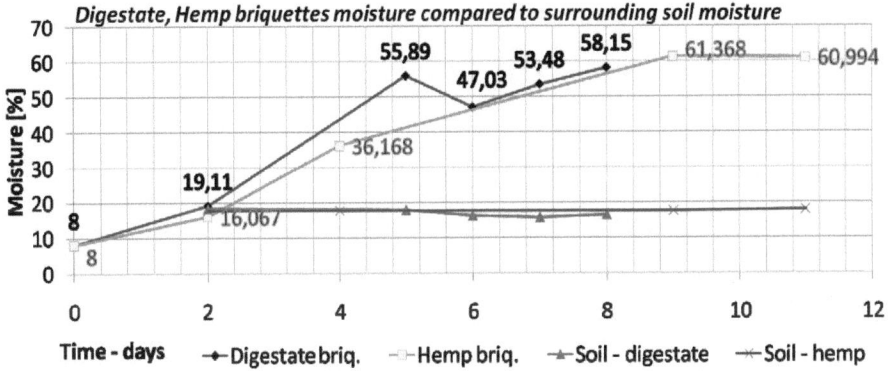

Figure 34: Moisture increase within the time, digestate and soil

From the initial value of briquettes moisture of 8-10% has increased during 2 days to 20% at digestate and to 16% at hemp. After next two days the moisture increased to 56% at digestate and 36% at hemp, beside the moisture of the soil kept the level of 16-20%. The highest increase of sorption has been recorded in first 4 days at digestate and 8 days at hemp. The speed of sorption in this part of experiment is much lesser than in part of limited diameter, it is caused by limited water supply. The difference between the soil moisture and briquette moisture varies between 30%-40%. The decreasing tendencies in graph describe the ability of briquette desorption, the opposite and reversible process of sorption.

Increasing of briquettes proportions causes the pressure on the surrounding soil and deforming it. Density and volume are increasing during the time of fluent sorption that points where graph tendency decreases can be caused by higher sorption of surrounding soil (Figures 35, 36).

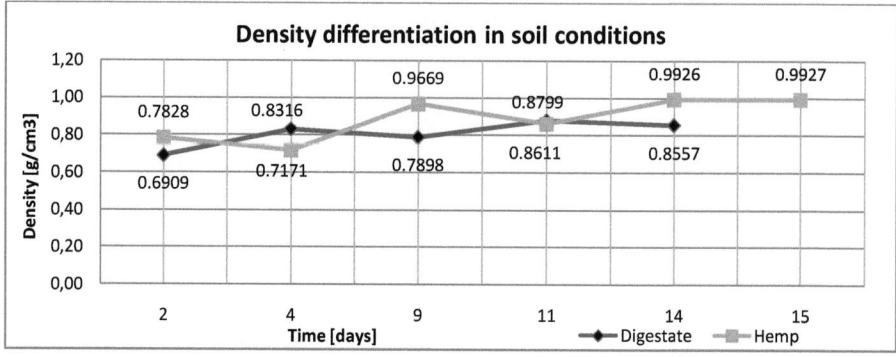

Figure 35: Density differentiation during the sorption

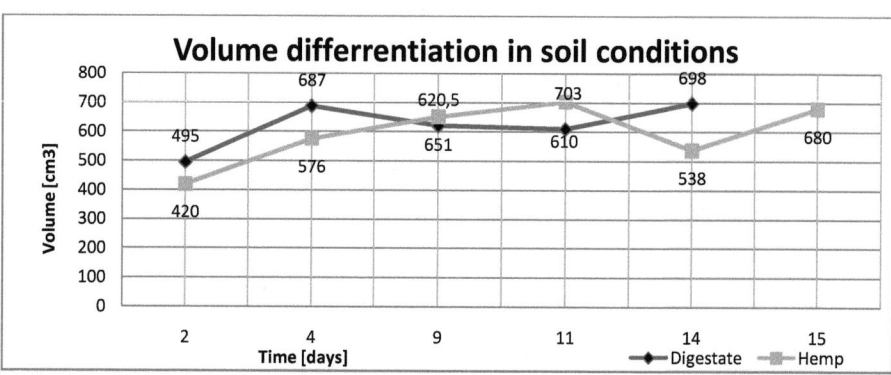

Figure 36: Volume differentiation during the sorption

6. CONCLUSIONS AND RECOMMENDATIONS

Studying of physico-mechanical properties of compressed digestate outlined the digestate good adsorptive properties, described by mass changes during the time of sorption, leading to increasing of its proportions and volume up to 5 times. The change in briquette volume leads to changes of mechanical properties and stability of briquette. Digestate briquettes compared to other material, like hemp or spruce bark showed that the composition of material influences the time of sorption and the stability of the briquette. Sorption is mainly influenced by the water supply, in cases of decreased access of water to briquette; the sorption lasted much longer, than in case of non-limited water supply. The digestate with mixture of additives, like zeolite, is easily compressed as well as pure digestate. By adding the additives, sorption was less caused by internal structure of crystalline material. The experiment of completely limited environment proved that digestate incorporated into the soil, has higher moisture than the surrounding soil. It means that digestate absorbs soil water from its environment and keep it in internal space, and soil maintains its moisture content in line by moving water molecules from further soil space. By the sorption of water the volume and the density is changed and there is assumption of complete decomposition in the soil with the prediction of two years.

The addition of mineral fertilizers does not significantly affect the mechanical properties of the obtained briquettes. Thanks to rich nutrient composition, digestate is suitable material to compressing process (compression with a diameter Ø60 mm is suitable only for digestate with humidity below 14%) and utilization in agriculture as an amendment material.

Obtained knowledge about sorption properties of digestate is one parts of the complex of properties, which could be used to its sustainable utilization in agriculture. Based on next research of this problematic these outlines should be confirmed. This study is not concerned to economical profitability and market sustainability of digestate processing and utilization in briquette form, thus I recommend to investigate even this part of the issue. Next research could be concerned to investigation of utilization the digestate briquettes in drier regions, based on third part of this study, when briquettes incorporated into the soil are keeping the moisture in its briquette structure.

7. REFERENCES

Abubaker, J., Risberg, K. and Pell, M. 2012. Biogas residues as fertilisers – Effects on wheat growth and soil microbial activities. *Applied Energy.* 2012. Vol. 99, pp. 126–134.

Ahlgren, S., et al. 2010. Nitrogen fertiliser production based on biogas – Energy input, environmental impact and land use. *Bioresource Technology.* 2010. Vol. 101, pp. 7181–7184.

Alburquerque, J. A., Fuente, C. de la and Bernal, M. P. 2012. Chemical properties of anaerobic digestates affecting C and N dynamics in amended soils. *Agriculture, Ecosystems and Environment.* 2012. 160, pp. 15-22.

Alburquerque, J.A., et al. 2012. Agricultural use of digestate for horticultural crop production and improvement of soil properties. *European Journal of Agronomy.* 2012. Vol. 43, pp. 119-128.

Alexander, R. 2002. Digestate utilization in the U.S. *BioCycle.* 2002. Vol. 53, p. 56. Available at www.biocycle.net/2012/01/digestate-utilization-in-the-us/.

Appels, L., et al. 2008. Principles and potential of the anaerobic digestion of waste-activated sludge. *Progress in Energy and Combustion Science.* 2008, 34, pp. 755–781.

BC Ministry of Environment. 2010. On-farm Anaerobic Digestion. *Waste Discharge Authorization Guideline.* 2010. Available at http://www.env.gov.bc.ca/epd/industrial/agriculture/pdf/anaerobic-digestion-guideline.pdf.

BIOCLEAN. 2013. Zeolit pro zemědělské užití. *Kámen zeolit .* [Online] BIOCLEAN , 2013. http://kamenzeolit.cz/zeolit-primes-do-substratu.

—. 2013. Zeolit pro zemědělské užití. *Kámen zeolit.* [Online] BIOCLEAN, 2013. [Cited: April 25, 2013.] Available at http://kamenzeolit.cz/zeolit-primes-do-substratu.

Bond, T. and Templeton, M. R. 2011. History and future of domestic biogas plants in the developing world. *Energy for Sustainable Development.* 2011, pp. 347–354.

Commission on colloid and surface chemistry. 1994. Recommendations for the characterization of porous solids. *International Union of Pure.* 8, 1994, Vol. 66, pp. 1739-1758. Available at http://www.oktatas.ch.bme.hu/oktatas/konyvek/fizkem/felfizkem/IUPAC_Sing.pdf.

ČSN ISO 1928. 2010. Tuhá paliva–Stanovení spalného tepla kalorimetrickou metodou v tlakové nádobě a výpočet výhřevnosti. Praha : Centrum technické normalizace TEKO, 2010. p. 24.

DEFRA. 2010. Fertiliser Manual. *(RB209).* s.l. : TSO - The Stationery Office, 2010. 8, p. 257. ISBN 978 0 11 243286 9.

Deublein, D. and Steinhauser, A. 2008. Biogas from Waste and. s.l. : Wiley-VCH, 2008. ISBN 978-3-527-31841-4.

EPA. 2012. Case Study Primer for Participant Discussion: Biodigesters and Biogas. *Technology Market Summit.* Washington D.C. : U.S. Environmental Protection Agency, May 14, 2012. p. 30. EPA 190S12005.

FAO. 2005. The importance of soil organic matter: Key to drought-resistant soil and sustained food and production. *FAO Soil Bulletin.* Rome : Food and Agriculture Organisation of the UN, 2005. 80. ISBN 92-5-105366-9.

Fixen, P. E. and West, F. B. 2002. Nitrogen Fertilizers: Meeting Contemporary Challenges. *A Journal of the Human Environment.* 2002, Vol. 31(2), pp. 169-176. Available at http://www.bioone.org/doi/full/10.1579/0044-7447-31.2.169.

Fletcher, A.J. 2008. Porosity and sorption behaviour. *Dr. Ashleigh Fletcher of the Department of Chemical and Process Engineering, University of Strathclyde.*

[Online] 2008. [Cited: April 20, 2013.] Available at http://personal.strath.ac.uk/ashleigh.fletcher/index.htm.

Frischmann, P. 2012. Enhancement and treatment of digestates from anaerobic digestion. *Desk top study on digestate enhancement and treatment.* s.l. : WRAP, 2012. Available at www.wrap.org.uk.

Gerardi, M. H. 2003. The Microbiology of Anaerobic Digesters. *Wastewater Microbiology Series.* New Jersey : A John Wiley & Sons, Inc., Publication, 2003. ISBN 0-471-20693-8.

Gómez, X., et al. 2005. Evaluation of digestate stability from anaerobic process by thermogravimetric analysis. *Thermochimica Acta.* 2005. Vol. 426, pp. 179–184.

Gong, W., Li, W. and Liang, H. 2010. Application of A/O-MBR for treatment. *Journal of Chemical Technology and Biotechnology.* April 30, 2010, 85, pp. 1334–1339.

Gruszkiewicz, M.S., et al. 2005. Water adsorption and desorption on microporous solids at elevated temperature. *Journal of Thermal Analysis and Calorimetry.* 2005. 81, pp. 609–615.

GTOS. 2009. Biomass. *Assessment of the status of the development of the standards for the Terrestrial Essential Climate Variables.* Rome : FAO, 2009. 10.

Chaudhary, B. K. 2008. Dry continuous anaerobic digestion of municipal solid waste in thermophilic conditions. Thailand : Asian Institute of Technology, School of Environment, Resources and Development, 2008. p. 93.

CHP Focus. 2013. Fuel Calorific Value. *Department of energy and climate change.* [Online] CHP Focus, 2013. [Cited: April 25, 2013.] http://chp.decc.gov.uk/cms/fuel-calorific-value/.

—. **2013.** Fuel Calorific Value. *Department of energy and climate change.* [Online] CHP Focus, 2013. [Cited: April 25, 2013.] Available at http://chp.decc.gov.uk/cms/fuel-calorific-value/.

Kaneko, K. 1994. Determination of pore size and pore size distribution 1. Adsorbents and catalysts. 96 *Journal of Membrane Science.* [PDF]. 1994. pp. 59-89.

Kratzeisen, M., et al. 2010. Applicability of biogas digestate as solid fuel. *Fuel.* 2010. Vol. 89, pp. 2544–2548.

Kumar, Sunil. 2012. BIOGAS. Rijeka, Croatia : InTech, 2012. 1. ISBN 978-953-51-0204-5.

Kužel, S., et al. 2010. Jak efektivně využít digestát. *Energie 21.* 2010, Vol. 3/10. Available at http://www.energie21.cz/archiv-novinek/Jak-efektivne-vyuzit-digestat__s303x46878.html.

Lukehurst, C.D., Frost, P. and Seadi, T. Al. 2010. Utilisation of digestate from biogas plants as biofertilizer. *IEA Bioenergy.* 2010.

Maier-Laxhuber, P. et al. 1991. Zeolite/Water Adsorption Cooling/Heating. *Rexresearch.com.* [Online] ZEO-TECH, 1991. [Cited: April 25, 2013.] Available at http://www.rexresearch.com/zeolite/maier.htm.

Makádi, M., Tomócsik, A. and Orosz, V. 2012. Digestate: A New Nutrient Source – Review. *InTech Europe.* s.l. : InTech, 2012. Available at: http://www.intechopen.com/books/biogas/digestate-a-new-nutrient-source-review. ISBN: 978-953-51-0204-5.

Mankowski, J. and Kolodziej, J. 2008. Increasing Heat of Combustion of Briquettes Made of Hemp Shives. *International Conference on Flax and Other Bast Plants.* 2008. pp. 344-352. ISBN 978-0-9809664-0-4.

Marešová, J. and Verner, L. 2012. Systém výroby a užití granulovaných paliv Ekover z fytomasy. *Ekover.* [Online] April 2012. [Cited: April 17, 2013.] Available at: http://www.ekover.cz/wp-content/uploads/2012/04/Ekover-a-%C3%BArodnost-p%C5%AFdy-Ek-a-odp.teplo-z-BS.pdf.

Mendelova zemědělská a lesnická univerzita v Brně. 2008. Příručka pro nakládání s digestátem a fugátem. Brno : s.n., 2008. 2, p. 30.

Monnet, Fabien. 2003. An Introduction to Anaerobic Digestion of Organic Wastes. *Final Report.* Scotland : s.n., November 2003. p. 48.

Paavola, T. and Rintala, J. 2008. Effects of storage on characteristics and hygienic quality of digestates from four co-digestion concepts of manure and biowaste. *Bioresour. Technol.* 2008. Vol. 99, 15, pp. 7041-7050.

Pechoušek, J. 2010. Měření plochy povrchu pevných látek a určování jejich porozity metodou sorpce. Olomouc : Univerzita Palackého v Olomouci, 2010. p. 19.

Piwoni, M.D. and Keeley, J.W. 1990. Ground Water Issue. *Basic Concepts of Contaminant.* s.l. : U.S. Environmental Protection Agenci, Oct 1990. Available at http://www.epa.gov/superfund/remedytech/tsp/download/lnapl.pdf.

Plöch, M. and Heiermann, M. 2006. Biogas farming in central and northern Europe: a strategy for developing countries? Invited overview. *Agriculture Engineering International.* 2006. Vol. 8, 8, pp. 1-15.

Ramos, M., Espinoza, R. L. and Horn, M.J. 2003. Evaluation of a zeolite-water solar adsorption refrigerator. *ISES Solar World Congress.* Göteborg : s.n., 2003. Available at http://fc.uni.edu.pe/mhorn/ISES2003%20(solar%20refrigeration).pdf.

READE Advance materials. 1997. Natural & Synthetic Zeolites. [Online] 1997. [Cited: April 25, 2013.] Available at http://www.reade.com/Products/Minerals_and_Ores/zeolites.html.

Richter, R. 2004. Sorpční schopnost půdy. *Ústav agrochemie a výživy rostlin, MZLU v Brně.* [Online] MZLU, 2004. [Cited: April 20, 2013.] Available:

http://web2.mendelu.cz/af_221_multitext/vyziva_rostlin/html/agrochemie_pudy/sorp ce.htm.

ROADEX. 2011. 2. Water in road materials and subgrade soils, terminology. *ROADEX.* [Online] 2011. [Cited: April 20, 2013.] Available: http://www.roadex.org/index.php/drainage2#2.6.

Seadi, Teodorita Al. 2001. Good practice in Quality Management of AD Residues. *IEA Bioenergy.* Denmark : University of Southern Denmark, 2001.

Seadi, Teodorita Al, et al. 2008. biogas HANDBOOK. s.l., Denmark : University of Southern Denmark Esbjerg, 2008.

Schovánek, P. and Havránek, V. 2011. Chyby a nejistoty měření. *Moderní technologie ve studiu aplikované fyziky.* 2011. p. 18. Available at: http://fyzika.upol.cz/cs/system/files/download/vujtek/texty/pext2-nejistoty.pdf.

Teglia, C., Tremier, A. and Martel, J.-L. 2011. Characterization of Solid Digestates: Part 1, Review of Existing. *Waste Biomass Valor.* Springer Science, 2011, 2, pp. 43-58.

Thomas, J. 2013. Mechanical Properties of Dolomitic Limestone. *eHow home.* [Online] 2013. [Cited: April 25, 2013.] http://www.ehow.com/list_6902102_mechanical-properties-dolomitic-limestone.html.

Volz, C. 2009. Lecture 5, Chemical Partitioning to Solids and Fugacity. *Center for Healthy Environments & Communities (CHEC).* [Online] 2009. [Cited: April 20, 2013.] Available: http://www.chec.pitt.edu/Academic%20Courses.html.

VŠCHT. 2008. Stanovení hustoty slitin - návod. [Online] Vysoká škola chemicko-technologická v Praze, 2008. [Cited: April 25, 2013.] http://www.vscht.cz/met/stranky/vyuka/labcv/labor/fm_termicka_analyza/hustota.ht m.

—. **2008.** Stanovení hustoty slitin - návod. [Online] Vysoká škola chemicko-technologická v Praze, 2008. [Cited: April 25, 2013.] Available at http://www.vscht.cz/met/stranky/vyuka/labcv/labor/fm_termicka_analyza/hustota.ht m.

VŠCHT, Ústav energetiky. 2012. Tuhá biopaliva z biomasy. *Vysoká škola chemicko technická.* [Online] 2012. [Cited: 4 17, 2013.] Available at: http://www.vscht.cz/ktt/studium/predmety/AZE_I/4AZE_I_2012.pdf.

Zaher, U., et al. 2007. Producing Energy and Fertilizer From Organic Municipal Solid Waste. *Ecology Publication No. 07-07-024.* 2007. Available at: http://www.ecy.wa.gov/programs/swfa/solidwastedata/.

8. LIST OF ILLUSTRATIONS

8.1 Tables

8.2 Figures

ANNEXES

Table 14: Mass measurements for all used materials

Mass [g]

Time [minutes]	Dig m1	Dig m2	Dig m3	Dig m4	Dig m5	Hemp m1	Hemp m2	SBark m1	SBark m2	Sorg p m1	Sorg p m2	Sorg PIChips m1	Sorg PIChips m2	Sorg Chip m1	Sorg Chip m2	Dig-lime m1	Dig-lime m2	Dig-lime m3	Dig-Zeolite m1	Dig-Zeolite m2
	n = 5					n = 2		n = 2		n = 2		n = 2		n = 2		n = 3			n = 2	
0	147,8	122,7	139,8	116,9	100,5	151,1	153,0	119,6	133,0	147,9	136,1	160,2	171,0	175,2	165,6	235,4	169,0	159,8	166,5	132,1
3	186,9	145,0	156,7	152,3	139,3	174,8	232,9	122,9	144,3	200,9	192,4	232,7	226,7	276,5	308,6	265,9	190,0	214,7	188,6	163,7
9	215,7	159,7	170,1	208,8	178,8	207,7	294,4	132,5	161,8	219,9	224,7	290,1	257,2	360,9	396,6	284,5	203,1	271,2	206,6	210,5
15	237,5	169,2	186,9	260,0	212,0	239,8	343,9	141,6	180,8	236,4	249,8	361,1	292,2	430,7	462,1	299,5	215,7	324,9	219,9	247,0
21	254,0	181,4	205,0	311,8	250,9	274,2	387,7	155,0	196,5	252,8	273,4	441,8	342,4	478,0	556,2	319,8	226,6	360,3	239,2	285,0
27	272,2	190,3	225,6	360,6	289,2	302,8	421,9	160,9	210,9	270,3	296,6	519,6	412,4			341,3	239,6	385,3	263,6	319,3
33	292,8	199,6	246,9	398,5	338,7	329,9	444,8	173,0	223,2	287,1	320,4	563,5	484,9			363,2	255,8	406,0	279,7	346,0
39	315,9	208,4	269,8	425,2	379,2	364,6	460,0	179,8	236,8	304,1	342,5	579,2	554,0			386,2	274,2	429,4	300,3	370,9
45	337,0	217,4	290,9	453,2	413,8	409,8	482,4	186,4	237,3	320,6	364,1	591,9	587,8			408,2	294,6	448,7	315,5	387,4
51	357,9	227,1	318,2	478,7	451,1	442,9	503,5	193,8	244,6	338,2	387,4	596,9	597,4			429,6	319,9	468,0	331,3	402,0
57	383,4	240,2	355,1	499,8	480,3	476,8	524,8	197,5	248,1	357,7	409,2	600,6	603,8			451,7	341,1	486,2	349,4	415,1
63	405,8	251,0	401,5	518,6	494,3	490,4	541,2	204,8	251,6							481,0	362,0	493,7	369,0	429,2
69	427,6	263,9	444,7	532,6	503,2	514,0	555,4	219,6	272,3							519,3	397,7	494,5	394,5	440,2
75	445,4	278,0	487,5	543,2	543,2	524,2	574,3	219,7	282,6							578,9	447,8	493,0	439,1	445,2
81	460,4	295,6	515,2	555,0	514,0	524,1	595,2	229,5	292,8							616,5	493,6		506,3	479,8
87	476,7	320,2	543,0	564,5	516,4	545,6	606,9	233,1	295,8							642,0	539,6		551,0	472,4
93	496,0	340,1	567,6	570,1	518,1	550,9	614,2	235,2	295,4							660,9	572,3		588,6	479,1
99	518,9	367,6	591,7	573,4	518,0	550,7	620,8	233,6	294,0							675,7	592,5		611,6	492,9
105	538,0	393,8	608,0	574,2	518,0	552,2	624,1	233,7	295,4							683,8	599,3		621,6	505,1
111	561,5	419,3	630,0	578,8												684,8	607,0		630,9	512,0
117	599,5	446,2	638,3	579,8												688,8	610,1		639,0	524,6
123	647,4	471,1	647,4	578,1												693,2	606,0		640,0	535,7
129	674,2	501,7	653,1													690,2			644,3	541,8
135	706,0	529,8	657,9													690,0			646,7	550,7
141	718,9	567,7	656,2																644,0	552,8
147	725,7	606,9																		562,9
153	727,4	611,5																		559,9
159	720,6	618,0																		564,9
165	724,6	613,8																		565,2

Table 15: Lenght measurements for all used materials

Lenght [cm]

Time [minutes]	n=5					n=2		n=2		n=2		n=2		n=2		n=3			n=2	
	Dig m1	Dig m2	Dig m3	Dig m4	Dig m5	Hemp m1	Hemp m2	SBark m1	SBark m2	Sorg p m1	Sorg p m2	Sorg PlChips m1	Sorg PlChips m2	Sorg Chip m1	Sorg Chip m2	Dig-lime m1	Dig-lime m2	Dig-lime m3	Dig-Zeolite m1	Dig-Zeolite m2
0	5,54	5,28	5,45	3,91	3,93	4,82	4,58	5,47	6,58	6,15	6,03	5,48	5,36	7,53	7,85	5,50	4,35	3,57	4,73	3,57
3	6,40	6,06	5,59	4,29	5,03	5,62	6,82	5,92	7,36	7,07	7,18	5,83	5,41	8,86	9,80	5,98	4,76	5,09	4,97	4,42
9	6,75	6,50	6,07	4,70	5,30	6,55	7,55	6,33	7,94	7,36	7,41	6,36	5,71	9,91	10,83	6,36	5,13	5,85	5,20	4,97
15	7,03	7,12	6,14	4,86	5,96	6,88	8,39	6,65	8,57	7,57	7,73	7,08	6,21	10,36	11,52	6,56	5,28	6,79	5,40	5,60
21	7,11	7,43	6,45	5,20	6,58	7,13	8,90	6,66	9,36	7,88	8,08	7,83	6,94	10,56	12,30	6,81	5,42	7,70	5,88	6,13
27	7,42	8,31	6,73	5,35	7,27	7,86	9,93	7,07	9,62	8,00	8,39	8,73	7,63			6,81	6,11	7,93	6,24	6,65
33	7,76	8,82	6,99	5,59	8,19	8,23	9,97	7,75	10,03	8,08	8,69	9,41	8,49			7,48	6,15	8,54	6,66	7,18
39	8,00	8,97	7,34	5,91	8,88	8,76	10,44	7,63	10,75	8,31	8,83	9,43	8,86			7,81	6,45	8,81	6,81	7,44
45	8,09	9,31	7,42	6,33	9,66	9,64	10,87	7,67	10,90	8,58	9,20	9,47	9,77			8,15	6,73	9,21	7,20	7,91
51	8,33	9,76	7,91	6,39	10,03	10,05	10,91	7,83	11,11	8,58	9,61	9,46	9,84			8,37	7,69	9,62	7,41	8,00
57	8,61	9,92	8,08	6,77	10,60					8,99	9,88	9,79	9,90			8,77	7,90	9,88	7,63	8,34
63	9,15	10,00	8,61	7,30	10,76											9,17	9,01	9,91	8,19	8,45
69	9,45	10,58	9,56	7,42	10,95											9,74	9,49	10,00	8,62	8,80
75	9,56	10,88	9,87	7,75	11,07											10,96	10,69	10,00	10,18	9,03
81	9,63	10,83	10,30	8,51												11,49	11,66		10,82	9,04
87	9,81	10,85	10,88	8,40												11,62	11,75		11,36	9,49
93	9,95	11,15	10,95	8,82												11,81	11,75		11,48	9,57
99	10,16	11,17	11,44	9,58												12,33	11,75		11,80	9,65
105	10,57	11,12	11,53	9,49												12,32	11,76		11,96	9,61
111	11,50		11,96													12,32			12,00	
117	11,66		11,92													12,32			12,07	
123	12,22		11,96													12,32			12,06	
129	12,49		11,93													12,32			12,02	

Table 16: Diameter measurements for all used materials

Diameter [cm]

Time [minutes]	n=5					n=2		n=2		n=2		n=2		n=2		n=3			n=2	
	Dig m1	Dig m2	Dig m3	Dig m4	Dig m5	Hemp m1	Hemp m2	SBark m1	SBark m2	Sorg p m1	Sorg p m2	Sorg P/Chips m1	Sorg P/Chips m2	Sorg Chip m1	Sorg Chip m2	Dig-lime m1	Dig-lime m2	Dig-lime m3	Dig-Zeolite m1	Dig-Zeolite m2
0	6,50	6,50	6,50	6,63	6,52	6,57	6,63	5,08	5,10	6,53	6,40	6,65	6,62	6,68	6,62	6,58	6,53	6,57	6,53	6,50
3	7,08	6,83	6,92	7,00	6,86	7,03	7,98	5,45	5,53	7,15	7,40	7,60	7,40	8,13	8,23	7,22	6,98	7,30	7,01	7,05
9	7,50	7,10	7,12	7,43	7,07	7,48	8,43	6,01	6,00	7,26	7,53	8,19	7,62	8,87	8,80	7,52	7,08	7,50	7,41	7,76
15	7,81	7,20	7,15	8,08	7,50	7,84	8,71	6,35	6,55	7,46	7,75	8,55	7,78	8,87	9,42	7,63	7,13	7,68	7,58	7,96
21	8,08	7,16	7,33	8,38	7,72	8,35	8,61	6,46	6,86	7,56	7,89	8,88	8,29	9,08	9,58	7,73	7,18	8,00	7,79	8,35
27	8,18	7,38	7,46	8,58	7,76	8,58	8,75	6,76	7,24	7,58	8,06	9,12	8,63			7,81	7,25	8,00	7,93	8,60
33	8,33	7,52	7,60	8,63	7,78	8,80	8,95	7,16	7,81	7,66	8,25	9,65	8,88			7,83	7,47	8,04	8,16	8,83
39	8,35	7,65	7,73	8,95	8,25	9,20	9,00	7,16	7,57	7,70	8,33	9,52	8,92			8,11	7,46	8,19	8,29	8,85
45	8,31	7,68	7,83	8,96	8,25	9,61	8,98	7,13	7,41	7,89	8,33	9,56	9,57			8,13	7,66	8,38	8,75	9,08
51	8,70	7,95	8,03	9,00	8,43	9,66	9,01	7,24	7,86	7,89	8,34	9,55	9,55			8,13	7,67	8,43	8,75	9,08
57	8,72	8,19	8,26	9,08	8,70					7,91	8,39	9,48	9,66			8,13	7,85	8,48	8,82	9,16
63	8,77	8,39	8,51	9,08	9,14											8,35	8,05	8,69	8,83	9,16
69	8,79	8,41	8,59	9,14	9,26											8,70	8,08	8,67	8,83	9,16
75	8,95	8,65	8,81	9,15	9,20											9,40	8,36	8,68	8,95	9,18
81	9,03	8,82	9,04	9,48												9,52	8,50		9,06	9,16
87	9,04	9,27	8,96	9,47												9,73	9,35		9,16	9,29
93	9,23	9,30	9,18	9,48												10,14	9,50		9,27	9,25
99	9,26	9,57	9,41	9,44												10,18	9,50		9,69	9,25
105	9,34	9,56	9,66	9,44												10,23	9,49		9,74	9,25
111	9,69		9,99													10,41			9,75	
117	9,81		10,06													10,41			10,06	
123	9,95		10,19													10,41			10,11	
129	9,95		10,14													10,41			10,12	

Table 17: Diameter limited conditions, mass and lenght increses, digestate and hemp

Time	n = 3				n= 3			
[hours]	Dig m1	Dig m2	Dig m3	Geomean	Hemp m1	Hemp m2	Hemp m3	Geomean
0:00:00	148,00	155,00	127,00	142,8	143,00	142,00	143,00	142,7
0:15:00	461,00	382,00	401,00	413,3	439,00	434,00	447,00	440,0
0:45:00	541,00	529,00	474,00	513,8	483,00	477,00	478,00	479,3
1:45:00	561,00	541,00	489,00	529,4	503,00	489,00	499,00	497,0
3:45:00	566,00	549,00	499,00	537,2	513,00	498,00	509,00	506,6
7:45:00	571,00	554,00	501,00	541,2	521,00	508,00	516,00	515,0

Time	n = 3				n = 3			
[hours]	Dig h1	Dig h2	Dig h3	Geomean	Hemp h1	Hemp h2	Hemp h3	Geomean
0:00:00	6,60	6,25	6,90	6,58	4,93	6,60	4,96	5,44
0:15:00	10,18	9,54	9,41	9,70	10,70	10,37	10,27	10,45
0:45:00	11,42	11,06	10,11	10,85	11,52	11,06	11,04	11,20
1:45:00	11,65	11,29	10,45	11,12	11,95	11,26	11,32	11,50
3:45:00	11,76	11,39	10,78	11,30	11,98	11,37	11,64	11,66
7:45:00	11,86	11,61	10,92	11,46	12,15	11,57	11,91	11,87

Table 18: Soil conditions, moisture measurement, digestate and hemp samples

Time	n = 2							n = 2		
[days]	w DigB1	w DigB2	Geomean	SD1	SD2	Volume	Density	w SoilD1	w SoilD2	Geomean
0	6,00	6,00	6,00	-	-	-	-	-	-	-
2	12,59	29,02	19,11	1,48	0,6	495	0,6909	18,53	18,24	18,38
5	55,11	56,68	55,89	0,85	1,33	687	0,8316	17,4	18,22	17,81
6	45,79	48,31	47,03	0,46	0,59	620,5	0,7898	16,13	16,33	16,23
7	49,6	57,67	53,48	0,24	0,54	610	0,8799	15,82	15,55	15,68
8	59,06	57,26	58,15	0,68	0,48	698	0,8557	16,65	16,19	16,42

Time	n = 2							n = 2		
[days]	w KonB1	w KonB2	Geomean	SD1	SD2	Volume	Density	w SoilK1	w SoilK2	Geomean
0	6,00	6,00	6,00	-	-	-	-	-	-	-
2	12,66	20,39	16,07	1,32	0,94	420	0,7828	17,78	17,77	17,77
4	22,4	58,4	36,17	1,7	1,22	576	0,7171	15,86	19,75	17,70
9	61,84	60,9	61,37	0,75	2,03	651	0,9669	17,69	17,31	17,50
11	60,12	61,88	60,99	2,01	1,35	703	0,8611	18,6	17,23	17,00
14	62,36	65,61	63,96	0,38	0,8	538	0,9926	19,21	15,22	17,10
15	66,3	51,2	58,26	1,03	1,16	680	0,9927	19,52	19,63	19,57

Printed by Books on Demand GmbH, Norderstedt / Germany